Helping Students Prepare for College Mathematics Placement Tests

A Guide for Teachers and Parents

Carmen M. Latterell

Rowman & Littlefield Education
Lanham, Maryland • Toronto • Plymouth, UK
2007

Published in the United States of America
by Rowman & Littlefield Education
A Division of Rowman & Littlefield Publishers, Inc.
A wholly owned subsidiary of The Rowman & Littlefield Publishing Group, Inc.
4501 Forbes Boulevard, Suite 200, Lanham, Maryland 20706
www.rowmaneducation.com

Estover Road
Plymouth PL6 7PY
United Kingdom

Copyright © 2007 by Carmen M. Latterell

All rights reserved. No part of this publication may be reproduced, stored in a retrieval system, or transmitted in any form or by any means, electronic, mechanical, photocopying, recording, or otherwise, without the prior permission of the publisher.

British Library Cataloguing in Publication Information Available

Library of Congress Cataloging-in-Publication Data

Latterell, Carmen M.
  Helping students prepare for college mathematics placement tests a guide for teachers and parents / Carmen M. Latterell.
    p. cm.
  Includes index.
  ISBN-13: 978-1-57886-655-7 (pbk. : alk. paper)
  ISBN-10: 1-57886-655-3 (pbk. : alk. paper)
  1. Mathematics—Study and teaching (Higher)—Problems, exercises, etc. 2. Mathematics—Study and teaching—Parent participation. 3. College entrance achievement tests—Problems, exercises, etc. 4. Mathematical readiness. 5. Education, Higher—Parent participation. I. Title.
  QA11.2.L368 2007
  510.76—dc22   2007016788

♾™ The paper used in this publication meets the minimum requirements of American National Standard for Information Sciences—Permanence of Paper for Printed Library Materials, ANSI/NISO Z39.48-1992.
Manufactured in the United States of America.

To my joy, my daughter, Lily Latterell

Contents

1	Introduction	1
2	Basic Skills without a Calculator	3
3	Graphing without a Graphing Calculator	13
4	A College Mathematics Placement Test	35
5	Elaborate Solutions to the Test	45
6	A Second College Placement Test with Solutions	67
7	Final Advice	85
Appendix 1: Study Skills for College Mathematics		87
Appendix 2: A Note to Teachers about State Standards		93
Index		95
About the Author		97

Chapter One
Introduction

College mathematics placement tests are becoming high stakes, just like all other testing. College freshmen take mathematics placement exams to determine their first mathematics course. Unfortunately, these tests tend to place students in courses lower than they should. At best, this adds on expenses and time to students' degree programs. At worst, it causes students to not succeed in the majors of their choice. Consequences in-between these extremes are also unpleasant, such as decreased chances at scholarships, awards, and external funding, because these types of things are reserved for students not taking remedial courses.

The issues with college mathematics placement tests have grown in number and complexity through the years. Part of the complexity is because secondary mathematics curricula do not align well with postsecondary mathematics curricula. The National Council of Teachers of Mathematics (NCTM) is a powerful organization with elementary teachers, secondary mathematics teachers, and mathematics educators (people who hold doctorates in mathematics education) among its members. The organization wrote a set of standards for K-12 mathematics in 1989 and revised it in 2000. These standards, commonly known as the NCTM Standards, called for radical changes in mathematics curricula in both content and pedagogy.

Mathematics educators created curricula to align with the NCTM Standards. These NCTM-oriented curricula have become popular and are in place throughout the United States. Research has shown many wonderful results from these curricula, and most mathematics educators believe that NCTM-oriented curricula are superior curricula to traditional curricula. In addition, these curricula align with state and national standards that schools must follow in order to receive state and national funding.

Most postsecondary mathematics departments are more traditional than they are NCTM-oriented. Mathematics placement tests are designed to place students into traditional courses (e.g., algebra, precalculus, and calculus). To see if students will succeed in these courses, professors want to see if students have working knowledge of procedures from the preceding courses. Thus, placement tests are made up of multiple-choice items that require the students to recall procedures and algorithms from traditional topics.

NCTM-oriented curricula emphasize mathematical processes such as problem solving. It is not that students from NCTM-oriented curricula lack procedural skills. It is only that procedural skills are not the emphasis of NCTM-oriented curricula. In addition, the procedures are spread over three or four years of a curriculum, and likely not reviewed in any depth in the senior year. Put together that NCTM-oriented curricula emphasize processes and placement tests emphasize procedures, and one can understand that under the pressure of a timed placement test, students place lower than they should.

Students also have a difficult time with college mathematics placement tests for reasons that reside with the tests and testing conditions. For example, these tests tend to consist of items that test nit-picky algebraic skills that students have either forgotten or never learned. These skills usually are not representative of students' overall mathematics ability. Further, students do not tend to prepare for these tests, and then do not perform up to their ability, due to nerves or simply running out of time.

This book is a guide through the material that students need to review and/or learn in order to avoid placing into remedial mathematics courses in college. The tone of the book is to explain concepts to a student, much as a teacher would do working with an individual student. Extremely motivated students can work their way through this book as a self-study guide. However, the easiest way to use this book is for secondary mathematics teachers and/or parents to use the book as a guide.

Although the tone of the book is to explain concepts that teachers and perhaps parents will already know, by using the book, teachers and parents will be spared trying to research what skills are required for college mathematics placement tests and at what level students need to understand these skills. The book also provides two full tests that students can actually take. In sum, this book contains everything teachers and parents need to prepare their students and/or children for college mathematics placement tests.

Chapter Two
Basic Skills without a Calculator

Upper-level college mathematics courses often take advantage of technology, including calculators and computers. However, in lower-level college courses, mathematics professors expect students to solve mathematics problems "by hand." This means, of course, without being dependent on a calculator. In this chapter, a set of basic concepts are reviewed. Most students already understand these concepts, but, it is surprising if a skill isn't used (such as adding decimals without a calculator), how easy it is to forget it. Parents and teachers can use the material that follows in this chapter to review these basic skills with students.

Negative Numbers

Some people claim that their trouble with mathematics began with the introduction of negative numbers. College students should certainly know how to add, subtract, multiply, and divide negative numbers. The rules are easy when multiplying or dividing. Simply multiply or divide as usual and leave the answer as a positive number if both original numbers were positive or both original numbers were negative. Leave the answer as a negative number if one of the original numbers was positive and one was negative.

-4 • -6 = 24
-4 • 6 = -24
4 • 6 = 24
4 • -6 = -24

When subtracting, change the problem to an addition. To do that, change the sign of the number that is being subtracted. For example, 5 - (-3) = 5 + (+3). The subtraction changed to addition, but then the negative 3 changed to a positive 3. Another example is -5 - 3 = -5 + -3. The subtraction changed to addition, but the positive 3 changed to a negative 3. In both cases, the first number remained unchanged.

When adding positive numbers, they are, of course, positive. The sum of two negative numbers is negative. And when adding one positive to one negative, subtract the smaller number (ignoring the sign) from the larger (ignoring the sign) and keep the sign of the larger number (ignoring the sign). A series of examples follow.

5 - -3 = 5 + 3 = 8
-5 - 3 = -5 + -3 = -8
5 + -3 = 2 (5 - 3 = 2, and keep the sign of the 5 which is larger than 3)
-5 + -3 = -8
5 + 3 = 8
-5 + 3 = -2 (5 - 3 = 2, and keep the sign of the 5 which is larger than 3)

Least Common Multiple (LCM)

Any positive integer can be written as a product of prime numbers. A prime number is a positive integer that has only two divisors (1 and the number itself). For example, the number 5 is prime and the number 4 is not prime (4 has divisors 1, 2, and 4). Some prime numbers follow: 2, 3, 5, 7, 11, 13, 17, 19, 23, 29, 31, 37, 41, 43, 47, 53, 59, 61, 67, 71, 73, 79, 83, 89, and 97.

A divisor is a positive integer that divides into the number without leaving a remainder. For example, the number 24 can be written as 4 • 6, but 4 and 6 are not prime. Yet, it is helpful to think of 24 as 4 • 6, and then break down 4 and 6. Four can be written as 2 • 2, and 6 can be written as 2 • 3. Thus, 24 can be written as 2 • 2 • 2 • 3, and those numbers are all prime.

Here is another example: 2520 can be written as 20 • 126. Twenty can be written as 2 • 10, and 10 can be written as 2 • 5. So far, we have 2 • 2 • 5, for the 20. Turning to the 126, it can be written as 6 • 21, and the 6 can be written as 2 • 3, and the 21 as 3 • 7. Gathering it all up, 2520 can be written as 2 • 2 • 2 • 3 • 3 • 5 • 7. One can check if this is correct by confirming that all the numbers are prime and that their product is 2520.

A least common multiple or LCM of a set of numbers is the smallest number out of the intersection of all multiples of the numbers. In other words, each number has an infinite number of multiples (generated by multiplying by 1, then 2, then 3, etc.). Collect all the multiples that are in common to all numbers and take the smallest. It is the LCM. But, there are easier methods to actually find one. Given two positive integers, their LCM can be found by first writing each number as its prime factorization.

To find the LCM of 24 and 2520, recall that 24 = 2 • 2 • 2 • 3 and 2520 = 2 • 2 • 2 • 3 • 3 • 5 • 7. The next step is to form a product out of the factors in the prime factorization of the separate numbers, but gather the numbers as many times as each occurs in *either* number. Thus, 2 occurs three times in 24 and three times in 2520, so gather three 2's. Three occurs once in 24 and twice in 2520, so gather two 3's. Five occurs zero times in 24 and once in 2520, so gather one 5, and also one 7, by the same reasoning. The LCM of 24 and 2520 is

2 • 2 • 2 • 3 • 3 • 5 • 7. It turns out that the least common multiple of 24 and 2520 is 2520. The smallest the LCM will be is the larger of the two numbers.

What is the LCM of 18 and 175? Eighteen can be written as 2 • 3 • 3, and 175 = 5 • 5 • 7. Since 2 and 3 do not occur in 175, and 5 and 7 do not occur in 18, these numbers have no factors in common. It turns out then that the LCM of 18 and 175 is 2 • 3 • 3 • 5 • 5 • 7, or 3150, which is 18 • 175. The product of the two numbers is the largest possible value for a LCM.

We have seen the LCM at its smallest and its largest. Sometimes the LCM is somewhere in-between. What is the LCM of 90 and 875? The prime factorization of 90 is 2 • 3 • 3 • 5 and of 875 is 5 • 5 • 5 • 7. Thus the LCM keeps a 2 (there is one 2 in 90 and none in 875, keep it the larger amount), keeps two 3's, keeps three 5's and one 7. The LCM is 2 • 3 • 3 • 5 • 5 • 5 • 7, or 15750. The product of 90 and 875 is 78750, which is, obviously, a larger number than the LCM. The LCM was smaller than the product of the two numbers because the two numbers had a 5 in common.

Fractions

Although a placement test might ask a direct question about a LCM, it is more likely that a placement test will ask a question about fractions. With fractions, the key skill to review is how to find a common denominator and LCMs are used in common denominators. One cannot directly add 1/4 and 1/8 without converting to the same denominator. A fraction is made up of two numbers, the top one called the numerator, and the bottom one called the denominator. To add or subtract fractions, the denominators must be the same number. If the denominators are not the same number, convert one or both denominators. The easiest denominator to use is the least common multiple, the LCM.

Once you determine the LCM of the denominators, it will be used as the new denominator. (16/90) + (20/875) can be computed by changing to fractions with a denominator of 15750, which is the LCM of 90 and 875. To convert 16/90, change the 90 to 15750 by multiplying by 175 (90 • 175 = 15750). Then multiply 16 by 175 (16 • 175 = 2800) and the new situation is 16/90 = 2800/15750. Similarly, 20/875 = 360/15750, because 875 • 18 = 15750 and 20 • 18 = 360. Now that the fractions have the same denominator, they can be added.

$$\frac{16}{90} + \frac{20}{875} = \frac{2800}{15750} + \frac{360}{15750} = \frac{3160}{15750}.$$

Some professors would expect the student to simplify that fraction, and others would not care. To simplify it, one cancels all common factors. For example, both the numerator and the denominator have a factor of 10.

To multiply fractions, just multiply the tops and multiply the bottoms. For example, (7/8)•(4/5) = 28/40, or, in general:

$$\frac{a}{b} \cdot \frac{c}{d} = \frac{ac}{bd}.$$

To divide fractions, remember the phrase "invert and multiply." In other words, division is turned into a multiplication problem; that is:

$$\frac{a}{b} \div \frac{c}{d} = \frac{a}{b} \cdot \frac{d}{c} = \frac{ad}{bc}.$$

For example, $(7/8) \div (4/5) = (7/8) \cdot (5/4) = 35/32$.

Exponents

A few rules are needed for working with exponents. First, an exponent, say a^n, means n a's are multiplied.

$$a^n = \underbrace{a \cdot a \cdot a \ldots a}_{n \text{ a's}}$$

The rules to be memorized include:

$$a^n \cdot a^m = \underbrace{a \cdot a \cdot a \ldots a}_{n \text{ a's}} \cdot \underbrace{a \cdot a \ldots a}_{m \text{ a's}} = a^{n+m},$$

when multiplying and the bases are the same, add the exponents.

$$\frac{a^n}{a^m} = a^{n-m},$$

when dividing and the bases are the same, subtract the exponents.

$$a^{-n} = \frac{1}{a^n},$$

a negative exponent reciprocates the base.

$$\left(a^n\right)^m = a^{n \cdot m},$$

when exponents are stacked, then multiply.

$$\sqrt[n]{a} = a^{\frac{1}{n}},$$

such as the square root is really an exponent of 1/2.

See the following examples.

$$3^4 = 3 \cdot 3 \cdot 3 \cdot 3$$
$$3^{25} \cdot 3^5 = 3^{30}$$
$$5^6 / 5^2 = 5^4$$
$$(2^7)^2 = 2^{14}$$
$$3^{-2} = 1/3^2 = 1/9$$

$2^2 \cdot 3^5$, other than literally calculating this, there is nothing to be done. For example, we could say $2^2 = 4$, and $3^5 = 234$, and so $2^2 \cdot 3^5 = 4 \cdot 243 = 972$. But, we cannot use exponent rules to combine the exponents in some manner, because the bases are not the same.

Decimals

A note of caution about decimals and collegiate mathematics. College mathematics professors are extremely precise people. If a student writes on a test or assignment that something is equal (=), it had better be equal and not approximately true. For example, $\pi = 3.14$ is a false statement. Pi is approximately 3.14, but it is not equal to 3.14. One could write $\pi \approx 3.14$, if wanted. But, even better, just use π as it is. Don't replace it with anything. Similarly, it is better to just use fractions as they are than to replace them with something that is only approximately true (such as a decimal). To belabor this point, it is true that $1/2 = .5$, but it is not true that $1/3 = .33$.

If stuck using decimals, students should know how to do arithmetic operations on them without a calculator. For addition and subtraction, line up the decimal points. $12.3 + 100.456$ can be done as follows.

$$\begin{array}{r} 12.300 \\ \underline{100.456} \\ 112.756 \end{array}$$

Whenever it is helpful, 0's can be placed to the right of a decimal point, after the non-zero numbers. Here is another: $123.456 + 45.03$

$$\begin{array}{r} 123.456 \\ \underline{45.030} \\ 168.486 \end{array}$$

Subtraction works the same way; that is, just line up the decimal points. Multiplication is also easy. Multiply the two numbers ignoring the decimal points altogether. Then count the total number of places to the right of the decimal in each number, and place the decimal point in the answer that same number of places from the right of the number.

For example, $123.456 \cdot 45.03$. First, multiply, ignoring the decimal points, to get 555922368. Now, count the total number of decimal spots to get 5 (three to the right in 123.456, and 2 to the right in 45.03). Thus, the answer is 5559.22368 which places the decimal point five spots from the right.

For division, I have always thought it was easiest to use common sense. For example, $112.34 \div 2.3$. Think of it as $112 \div 2$, which is about 50. Now, divide 11234 by 23 (I've ignored the decimals). This will result in 488 and a remainder. Since the answer must be around 50, the answer is 48.8. This is probably close enough, as long as it is marked as approximate. That is, $112.34 \div 2.3 \approx 48.8$.

What is $52.05 \div .006$? Think of the problem without decimals. What is $5205 \div 6$? Work that division to get 8675 (ignoring the decimal point). Now, the original problem was something like 50 divided by 5/1000. This is harder to think about than the previous one, but just quickly divide this (remember to invert and multiply) to get 10,000. The answer is less than 10,000 but not by a lot. The answer will be approximately 8675.

What is 1032 ÷ 45.678? Again, work this problem without the decimals, so what is 1032 ÷ 45678? The answer is 225929 (well, it has a decimal in front of it, but we can ignore that). Now, 1032 ÷ 45 is approximately 23. The answer would have to be approximately 22.59.

Complex Numbers

Complex numbers are of the form a + bi, where i is the square root of -1 (in other words, i^2 = -1) and a and b are real numbers. To add complex numbers, add the real parts and add the parts that go with the i. That is, a + bi added to c + di gives (a + c) + (b + d)i. Subtracting works the same way. a + bi subtracted from c + di gives (c - a) + (b - d)i. When multiplying complex numbers, just "foil." That is:

$$(a + bi)(c + di) = ac + adi + bic + bdi^2.$$

Recall that i^2 = -1. So, the product can be simplified to (ac - bd) + (ad + bc)i. When dividing complex numbers, multiply by the so-called conjugate of the denominator. The conjugate of a + bi is a – bi, and vice versa.

$$\frac{a+bi}{c+di} = \frac{a+bi}{c+di} \cdot \frac{c-di}{c-di} = \frac{ac - adi + bic - bdi^2}{c^2 - d^2 i^2}$$

This can be simplified by replacing i^2 with -1.

$$\frac{ac - adi + bic - bdi^2}{c^2 - d^2 i^2} = \frac{(ac+bd) + (bc-ad)i}{c^2 + d^2}$$

And, although it doesn't look very simple, the denominator is just a number. So, the entire thing can be written as one complex number:

$$\frac{(ac+bd) + (bc-ad)i}{c^2+d^2} = \frac{ac+bd}{c^2+d^2} + \frac{bc-ad}{c^2+d^2} i$$

Examples follow.

$$4 + 5i + 2 - 3i = (4+2) + (5-3)i = 6 + 2i$$

$$(2+3i) \cdot (7+4i) = 2\cdot7 + 2\cdot4i + 3i\cdot7 + 3i\cdot4i = 14 + 8i + 21i + 12\cdot-1 = 2 + 29i$$

$$\frac{2-3i}{8-6i} = \frac{2-3i}{8-6i} \cdot \frac{8+6i}{8+6i} = \frac{16 + 12i - 24i - 18i^2}{64 - 36i^2} = \frac{34 - 12i}{100} = \frac{34}{100} - \frac{12}{100} i$$

A Note about Square Roots

A square root of n equals the number whose square is n. That is,

$$\sqrt{n} = a, \text{ if } a^2 = n.$$

Sometimes professors are concerned that square roots are not left in the denominator of a fraction. For example, consider 1 divided by the square root of 2. To

Basic Skills without a Calculator

bring the square root of 2 out of the denominator, multiply the top and bottom by the square root of 2. Multiplying both the top and the bottom is the same as multiplying by 1; that is,

$$\frac{1}{\sqrt{2}} \cdot \frac{\sqrt{2}}{\sqrt{2}} = \frac{\sqrt{2}}{\sqrt{4}} = \frac{\sqrt{2}}{2}, \text{ because } \sqrt{a} \cdot \sqrt{a} = a.$$

Consider a more difficult situation,

$$\frac{x+3}{\sqrt{x-2}-3}.$$

This time multiply by

$$\frac{\sqrt{x-2}+3}{\sqrt{x-2}+3}.$$

The reason for switching the operation between the square root and the 3 from a subtraction to an addition is to avoid the middle term when multiplying.

$$\frac{x+3}{\sqrt{x-2}-3} \cdot \frac{\sqrt{x-2}+3}{\sqrt{x-2}+3} = \frac{(x+3) \cdot (\sqrt{x-2}+3)}{x-2-9}.$$

In the numerator, nothing has been done yet. In the denominator, the

$$\sqrt{x-2} \cdot \sqrt{x-2} = x-2.$$

There is no middle term, because it is the sum of

$$-3\sqrt{x-2} \text{ and } 3\sqrt{x-2},$$

which is 0. And the last term is -3 times 3, which is -9. The bottom simplifies to $x - 11$. The top is not so simple. It could be multiplied out. However, often the point is to remove the square root from the denominator, which is accomplished. So, perhaps one can just leave the whole thing as

$$\frac{(x+3) \cdot (\sqrt{x-2}+3)}{x-11}.$$

A Note about Sigma

The sigma notation means sum up. In college, it will often have indices. For example,

$$\sum_{i=1}^{5} 2i$$

has an index of 1 that ends at 5. The $i=1$ means start at 1 and increment by one until reaching 5. Two examples follow.

$$\sum_{i=1}^{5} 2i = 2 \cdot 1 + 2 \cdot 2 + 2 \cdot 3 + 2 \cdot 4 + 2 \cdot 5 = 30.$$

$$\sum_{i=1}^{2} (i^2 + 3) = 1^2 + 3 + 2^2 + 3 = 11.$$

Another problem, one that confuses a lot of students, is

$$\sum_{i=1}^{5} 2.$$

Sometimes students think the answer is 2, since there is no i in the function. But, i still increments, whether it is in the formula part or not. So,

$$\sum_{i=1}^{5} 2 = 2+2+2+2+2 = 10.$$

Matrix Arithmetic

A matrix is a rectangular array of numbers. It has rows and columns, with the size given as the number of rows by the number of columns. Matrix addition and subtraction are defined as long as the two matrices are the same size (i.e., same number of rows and same number of columns), in which case, entry by entry the entries are added or subtracted. For example,

$$\begin{bmatrix} 3 & 2 & -1 \\ 5 & 4 & 7 \\ -3 & 1 & 0 \end{bmatrix} + \begin{bmatrix} 1 & 6 & -3 \\ -2 & 0 & 8 \\ 3 & 11 & 5 \end{bmatrix} = \begin{bmatrix} 4 & 8 & -4 \\ 3 & 4 & 15 \\ 0 & 12 & 5 \end{bmatrix}$$

Matrix multiplication is not defined as multiply each entry by each entry. Rather, to be defined, if the first matrix is of size $n \times m$, then the second matrix must be of size $m \times r$. That is, the number of columns in the first matrix must equal the number of rows in the second matrix. Then to get entries, follow this rule:

$$c_{i,j} = a_{i,1} \cdot b_{1,j} + a_{i,2} \cdot b_{2,j} + a_{i,3} \cdot b_{3,j} + \ldots + a_{i,m} \cdot b_{m,j},$$

where the a entries come from the first matrix, the b entries from the second matrix, and the c entries are the product. Entries are numbered by row first, then column. For example, $a_{3,4}$ is an entry from the first matrix, third row, and fourth column. An example follows.

$$\begin{bmatrix} 2 & 4 \\ 6 & 4 \\ -1 & 7 \\ 9 & -3 \end{bmatrix} \cdot \begin{bmatrix} 1 & 4 & 3 \\ -2 & 0 & 5 \end{bmatrix} = \begin{bmatrix} 2\cdot1+4\cdot-2 & 2\cdot4+4\cdot0 & 2\cdot3+4\cdot5 \\ 6\cdot1+4\cdot-2 & 6\cdot4+4\cdot0 & 6\cdot3+4\cdot5 \\ -1\cdot1+7\cdot-2 & -1\cdot4+7\cdot0 & -1\cdot3+7\cdot5 \\ 9\cdot1+-3\cdot-2 & 9\cdot4+-3\cdot0 & 9\cdot3+-3\cdot5 \end{bmatrix}$$

$$= \begin{bmatrix} -6 & 8 & 26 \\ -2 & 24 & 38 \\ -15 & -4 & 32 \\ 15 & 36 & 12 \end{bmatrix}$$

Basic Skills without a Calculator

Matrix division doesn't exist as such. However, there is such a thing as a matrix inverse. A matrix inverse is the matrix one can multiply by to get the identity matrix. The identity matrix is a square matrix with 1's on its diagonal and 0's elsewhere. A matrix only has an inverse if it is square, and even then it might not have one. To find a matrix inverse takes some work. It is unlikely that such a problem would occur on a placement test. If it does, it would be for a small matrix, such as a 2 x 2 one. Take the matrix, and append the identity matrix. Then use row reduction techniques until the original matrix is replaced by the identity. The following example goes step by step. Given

$$\begin{bmatrix} 2 & 4 \\ 6 & 8 \end{bmatrix},$$

append the identity matrix.

$$\begin{bmatrix} 2 & 4 & 1 & 0 \\ 6 & 8 & 0 & 1 \end{bmatrix}.$$

The goal is to make the 2 a 1, the 4 a 0, the 6 a 0 and the 8 a 1. To change the 2 to a 1, divide by 2. If the first entry is divided by 2, then all entries in that row must be divided by 2. The matrix is now

$$\begin{bmatrix} 1 & 2 & \frac{1}{2} & 0 \\ 6 & 8 & 0 & 1 \end{bmatrix}.$$

Next, make the 6 a 0. To do that, multiply the top row by -6, and add it to the second row.

So, 1•-6 + 6 = 0
2•-6 + 8 = -4
(1/2)•-6 + 0 = -3.
And 0 • -6 + 1 = 1. So, the new matrix is

$$\begin{bmatrix} 1 & 2 & \frac{1}{2} & 0 \\ 0 & -4 & -3 & 1 \end{bmatrix}.$$

Next, the -4 needs to be a 1. Divide everything in that row by -4. This yields

$$\begin{bmatrix} 1 & 2 & \frac{1}{2} & 0 \\ 0 & 1 & \frac{3}{4} & \frac{-1}{4} \end{bmatrix}.$$

Finally, the 2 must be changed to a 0. Multiply the bottom row by -2 and add it to the top row.
So, 0•-2 + 1 = 1
1 •-2 +2 – 0
$\frac{3}{4}$•-2+$\frac{1}{2}$=-1

$\frac{-1}{4} \bullet -2 + 0 = \frac{1}{2}$. Thus, the matrix is

$$\begin{bmatrix} 1 & 0 & -1 & \frac{1}{2} \\ 0 & 1 & \frac{3}{4} & \frac{-1}{4} \end{bmatrix}.$$

That means the inverse is

$$\begin{bmatrix} -1 & \frac{1}{2} \\ \frac{3}{4} & \frac{-1}{4} \end{bmatrix}.$$

Chapter Three
Graphing without a Graphing Calculator

Most students actually do have the basic arithmetic skills discussed in the previous chapter. However, it is becoming more common for students to not have graphing skills. College mathematics professors expect that students can graph without a graphing calculator. One way to master graphing is to think in terms of function families.

Graphs of Lines

Some functions, such as $f(x) = 3x + 2$, are lines. Functions that are lines take the form $f(x) = mx + b$, where m is the slope of the line and (0, b) is a point on the line. To graph a line, plot any two points and connect them. In the case of $f(x) = mx + b$, use (0, b) as a point, and then pick any other point by picking an x-value and plugging it into the equation. For example, using $x = 3$, then $f(3) = 3 \cdot 3 + 2 = 11$, so another point is (3, 11).

Students can also think about the slope of the line. The slope of a line is the steepness of it. It is a fraction (even if it is a whole number, think of it as that whole number divided by one). Slope is rise over run. It can be positive or negative. I tell students to put the negative (if there is one) with the top number. To use slope to graph, use the top number to go up or down. Go down if the top number is negative and up if the top number is positive. Always go right, and go right the amount of the bottom number. For example, if one point on the graph is (2, 3) and the slope is 4, then move up four units from 3 and right one unit from 2. Thus, (3, 7) is a second point on the graph.

Given any two points on a line, just plot them and connect them. To find an equation for the line given two points, use $y - y_1 = m(x - x_1)$, where m is the slope of the line and (x_1, y_1) is any point on the line. Of course, one will have to solve to find the slope:

$$m = \frac{y_2 - y_1}{x_2 - x_1}.$$

For example, if (3, 4) and (10, 12) are two points, then the slope is:

$$m = \frac{y_2 - y_1}{x_2 - x_1} = \frac{12 - 4}{10 - 3} = \frac{8}{7}$$

and an equation for the line is $y - 12 = (8/7)(x - 10)$. Most professors will want students to rearrange that equation so it is of the form $f(x) = mx + b$. In this case, $y = (8/7)x - (80/7) + 12$ or $y = (8/7)x + (4/7)$.

When two lines are parallel to each other, their slopes are equal. If two lines are perpendicular to each other, their slopes are negative reciprocals. For example, if one has slope (a/b), then the other has slope (-b/a).

Graphs of Parabolas

Some functions are parabolas. They take the form $f(x) = ax^2 + bx + c$, where a, b, and c are numbers, but a cannot be 0. The simplest quadratic function (parabola) is $f(x) = x^2$. It opens upward and has a vertex at (0, 0). The graph of $f(x) = x^2$ is given below, and that is the general shape of any parabola.

Image 3.1: A Parabola

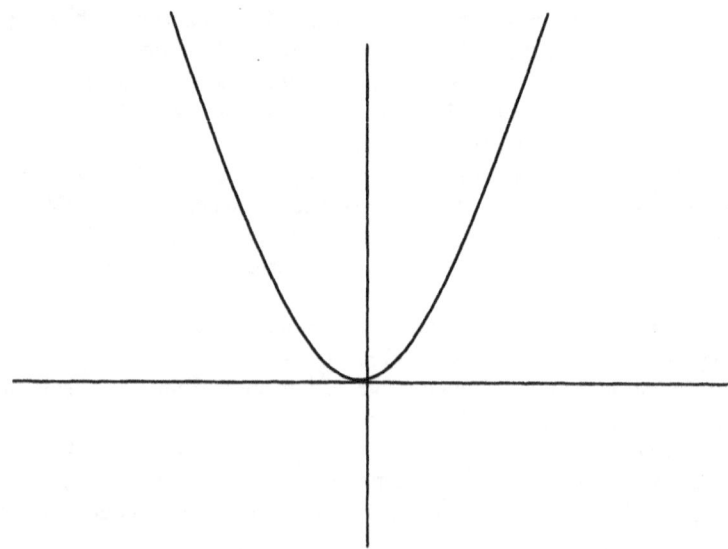

To graph a parabola, if the a is positive, it opens up, and if a is negative it opens down. The vertex is at (h, k) if the function is written in the form $f(x) = a(x - h)^2 + k$. Rewriting the function to the last form is not trivial, as one has to "complete the square". Students should know how to complete the square.

Given a quadratic expression $x^2 + bx$, to complete the square, one wants to write it as $(x + \text{something})^2$. That is, find something to add on so that the expression factors in this special way. The "something" to add on is $(b/2)^2$. For example, given $x^2 + 6x$, then add on $(6/2)^2 = 9$. Adding on 9 results in: $x^2 + 6x + 9 = (x + 3)(x + 3) = (x + 3)^2$, a perfect square.

If a parabola is in this form $f(x) = ax^2 + bx + c$, to pick out the vertex, write it in this form $f(x) = a(x - h)^2 + k$. First, the a coefficient needs to get out of the way, so do this: $f(x) = ax^2 + bx + c = a(x^2 + (b/a)x) + c$. For example, if the equation is $f(x) = 2x^2 - 6x + 7$, then $f(x) = 2x^2 - 6x + 7 = 2(x^2 - 3x) + 7$. Next, complete the square, so add on $(-3/2)^2 = 9/4$. But, we cannot just add on 9/4, as it will change the equation to a different one. It is fine, however, to add on 0. We can write 0 as $(9/4) - (9/4)$ or something similar that will help us. If 9/4 is added on inside the parentheses, that is, $2(x^2 - 3x + (9/4))$, then we have really added on 18/4 (because the 9/4 was multiplied by 2). We can write 0 as $18/4 - 18/4$. Let's see what this looks like:

$$f(x) = 2(x^2 - 3x + \frac{9}{4}) + 7 - \frac{18}{4} = 2(x - \frac{3}{2})^2 + \frac{10}{4}.$$

We went through all that work so that we know that the vertex is at (3/2, 10/4).

In calculus, students may have to complete the square from time to time. But, when graphing parabolas, it might be easier to do other things. For example, one could factor $ax^2 + bx + c$ to find the roots. These roots will be where the parabola crosses (or intercepts) the x-axis. This might give enough of an idea what the parabola looks like. One can always plot (0, c), the point on the parabola where it intercepts the y-axis.

Memorize These Three Graphs

There are a few other graphs (besides lines and parabolas) that students should be able to get almost instantly. For example, Image 3.2 contains the graphs of x-cubed and the square root function.

Image 3.2: $f(x) = x^3$, a cubic

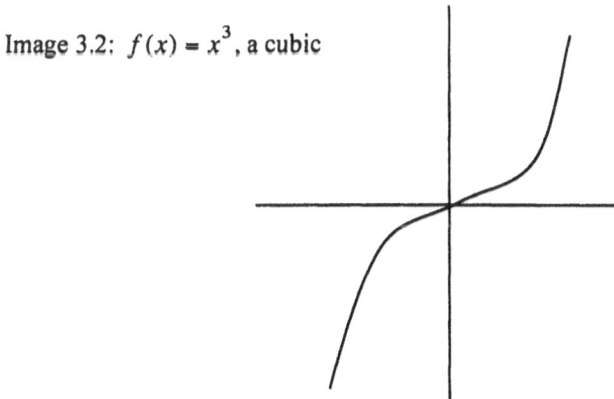

$f(x) = \sqrt{x}$, the square root function

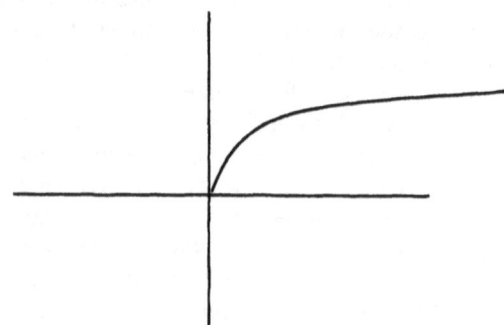

In addition, memorize the basic absolute value graph. The absolute value function is defined: $f(x) = x$, if $x \geq 0$ and $f(x) = -x$, if $x < 0$. That is, the absolute value of a positive number is simply that number, and the absolute value of a negative number is the negative of the number, because negating a negative makes it positive. For example, $|3| = 3$, and $|-3| = --3 = 3$, where the bars mean to take the absolute value. When graphing absolute value functions, they form V's. The basic absolute value function follows.

Image 3.3: $f(x) = |x|$

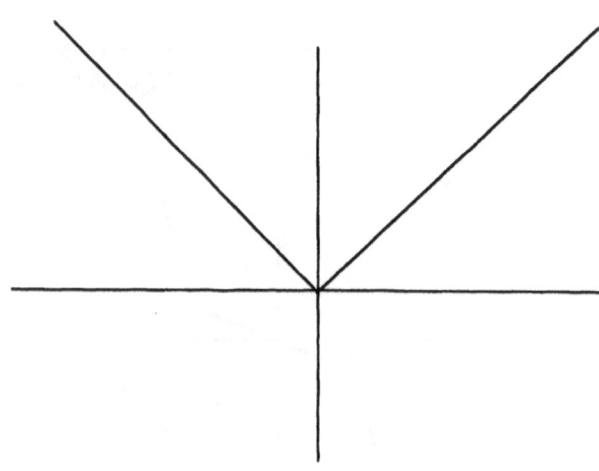

Exponential Graphs

Exponential functions are also important. Exponentials take the form of $f(x) = a^x$, where a is a positive real number. Below is a graph of $f(x) = 2^x$, and whenever a > 1 the graph will take the same form as this one, called exponential growth.

Image 3.4: $f(x) = 2^x$, an exponential growth function

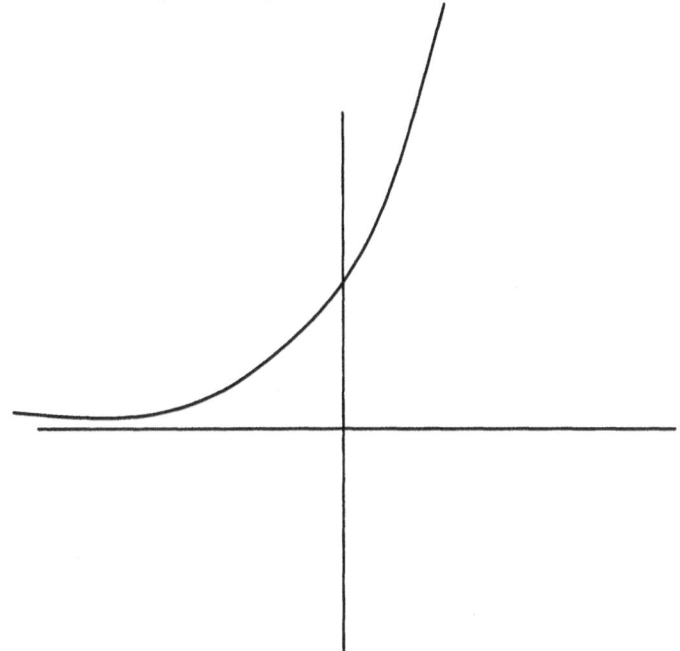

Image 3.5 is a graph of $f(x) = .5^x$, and whenever a is a number between 0 and 1, the graph will take the same form as $f(x) = .5^x$, called exponential decay.

By the way, when $a = 1$, the graph is not exponential, it is just a constant function. Exponential functions are easy to graph if you know the general shapes, and then plot two points (0, 1) and (1, a). In fact, these two points will ensure that the shape is going the right way. Exponential functions occur in many applications (e.g., compound interest, bacteria growth, and radioactive decay).

Graphing Rules

It is wise to keep some basic graphs in mind and be able to make some modifications.

Given a function, $f(x)$, and a positive value, a, then
- $f(x) + a$ raises $f(x)$ a units
- $f(x) - a$ lowers $f(x)$ a units

- $f(x + a)$ shifts $f(x)$ left a units
- $f(x - a)$ shifts $f(x)$ right a units.

Image 3.5: $f(x) = .5^x$, exponential decay

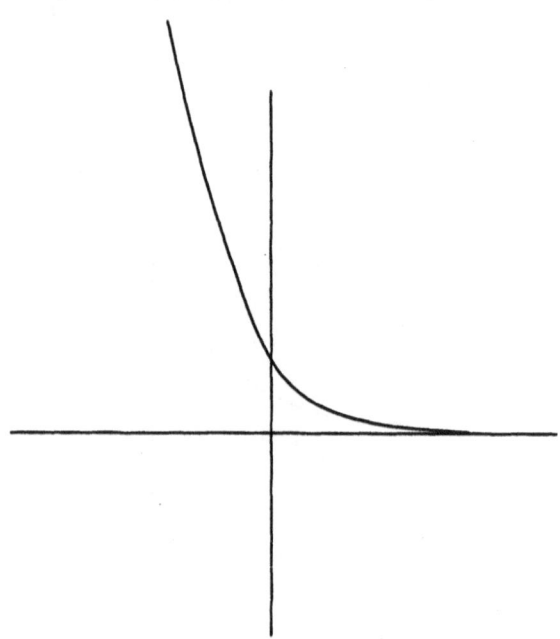

Given a function, $f(x)$, and a value, a, greater than 1, then
- $af(x)$ vertically stretches $f(x)$ by a factor of a
- $f(ax)$ horizontally compresses $f(x)$ by a factor of a.

Given a function, $f(x)$, and a value, a, with $0 < a < 1$, then
- $af(x)$ vertically compresses $f(x)$ by a factor of a
- $f(ax)$ horizontally stretches $f(x)$ by a factor of a.

Given a function, $f(x)$, and a value, a, equal to -1, then
- $af(x)$ reflects $f(x)$ about the x-axis
- $f(ax)$ reflects $f(x)$ about the y-axis.

For example, say that $f(x)$ is the following graph and then follow through the series of graphs given in Image 3.6.

Image 3.6: a) *f*(*x*), an arbitrary function

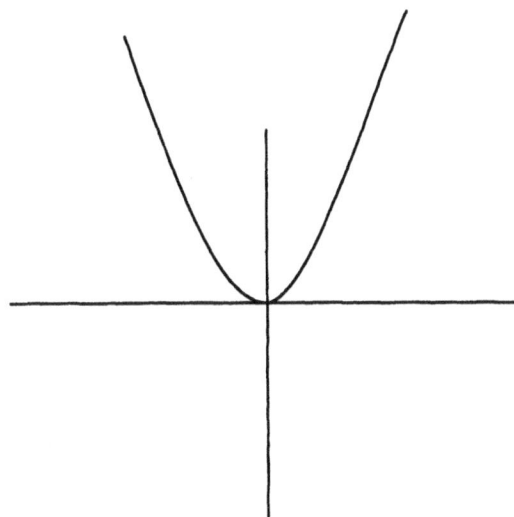

b) *f*(*x* + 2)

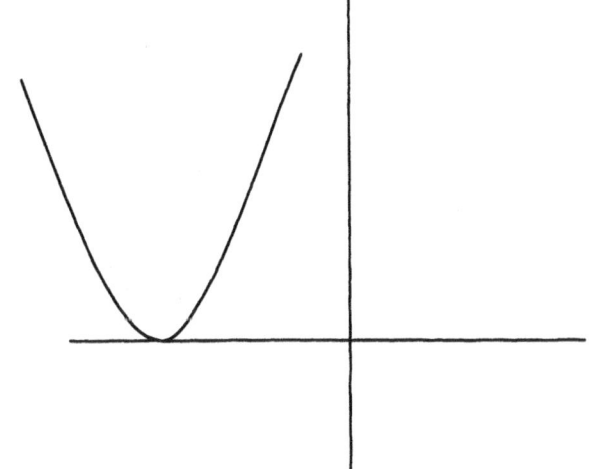

c) $f(x - 2)$

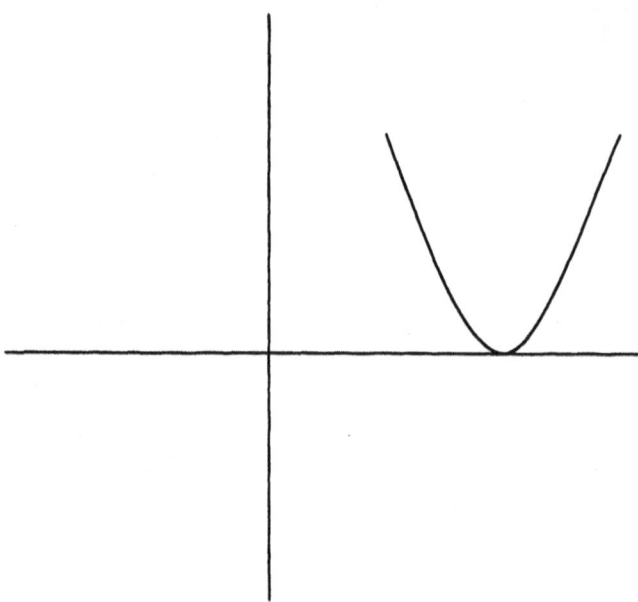

d) $f(x) + 2$

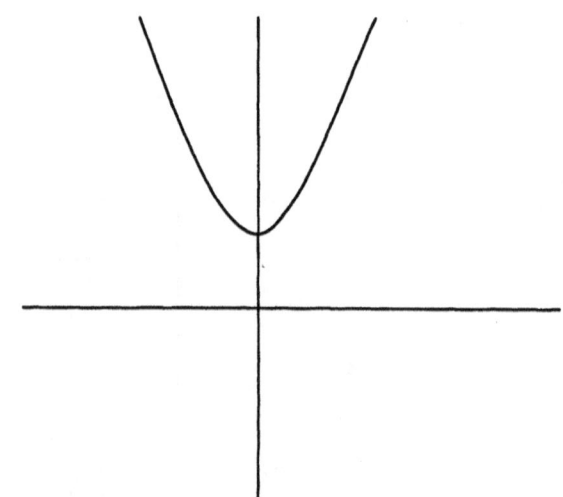

e) $f(x) - 2$

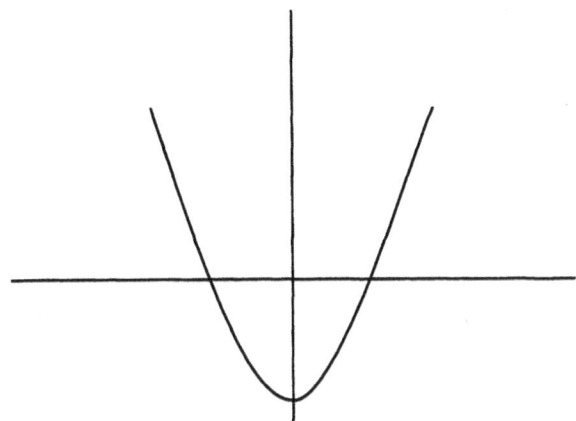

f) The inside graph is $2f(x)$ and the outside graph is $f(x)$.

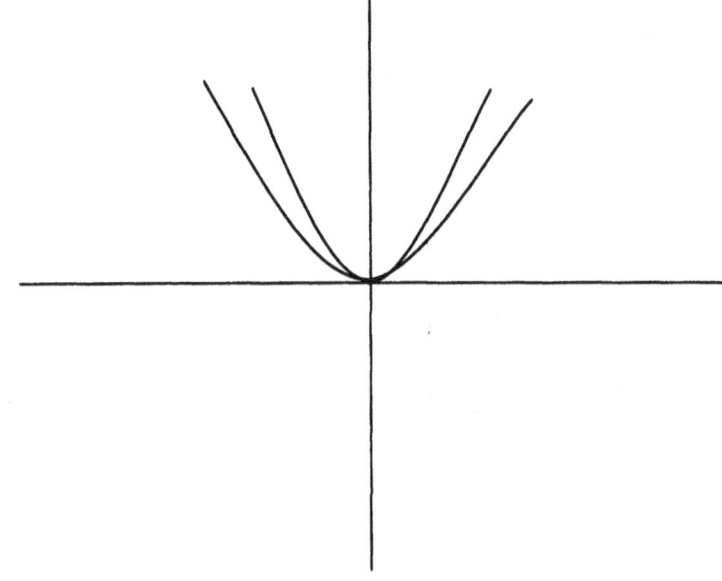

g) The outside graph is $\frac{1}{2}f(x)$ and the inside graph is $f(x)$.

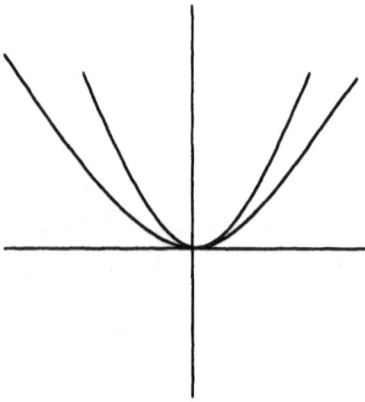

h) $-f(x)$

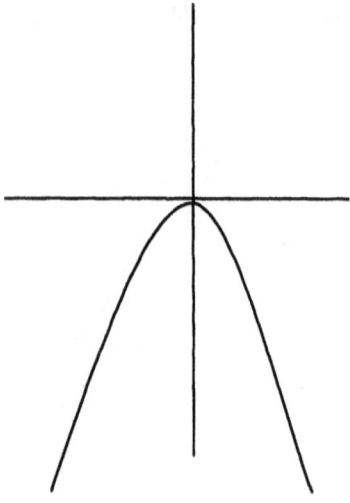

Graphing without a Graphing Calculator 23

i) The inside graph is $f(2x)$ and the outside graph is $f(x)$.

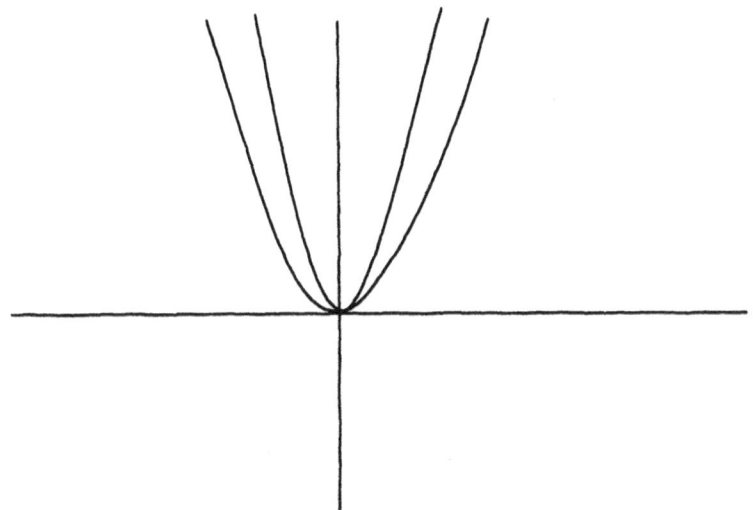

j) The outside graph is $f(.5x)$ and the inside graph is $f(x)$.

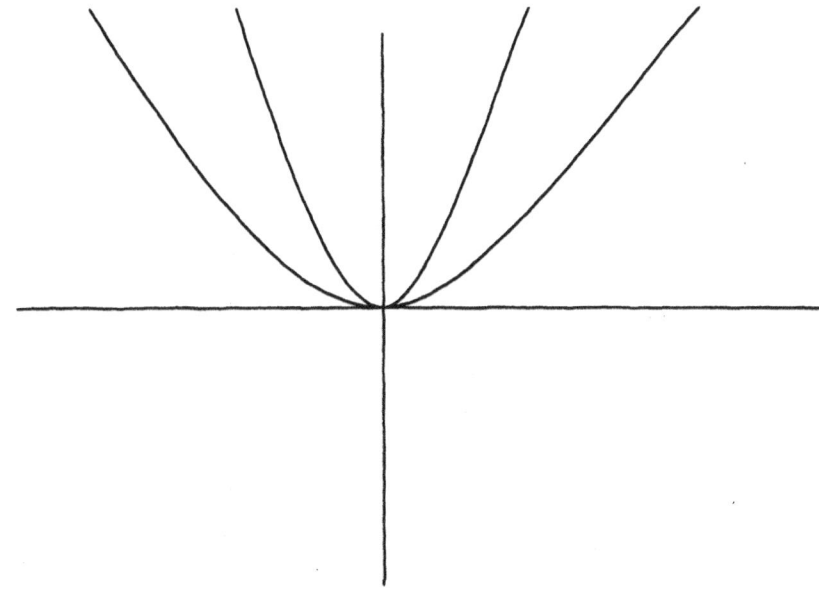

Notice that I have not given an example of $f(-x)$. That is because in my example $f(x) = f(-x)$. This sometimes happens. See Image 3.7 below for an example where $f(x) \neq f(-x)$.

Image 3.7: a) $f(x)$

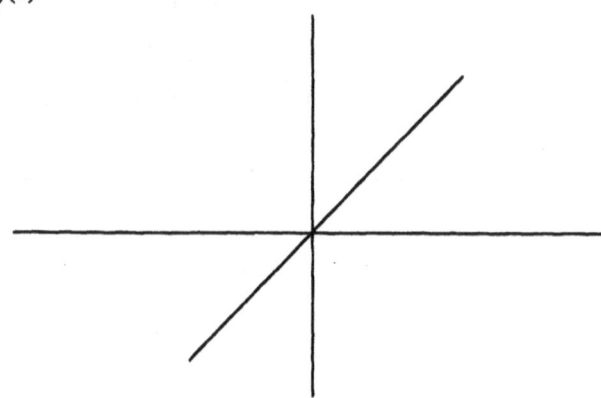

b) $f(-x)$

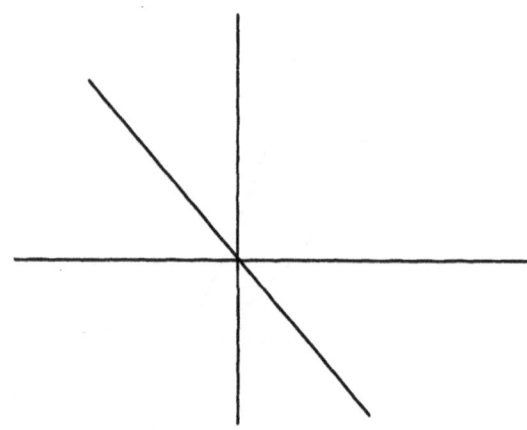

Graphs of Polynomials

Polynomials are continuous functions and easy to graph. First, to recognize a polynomial, the function will take a form like this: $f(x) = ax^n + bx^{n-1} + cx^{n-2} + \ldots$

Graphing without a Graphing Calculator

+ dx + e, where a, b, c, d, and e are numbers, $a \neq 0$. Polynomial functions are continuous with no breaks or gaps in their graphs. The graph of the function simply moves along in a smooth fashion and can switch directions. See the graph of the following polynomial.

Image 3.8: A Polynomial

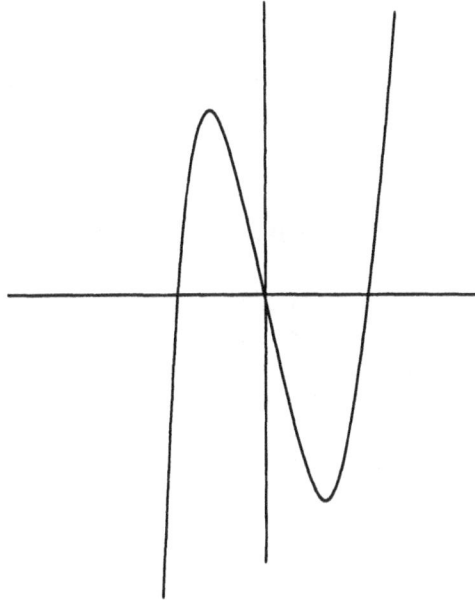

The easiest way to graph a polynomial is to factor it. Then set each of the factors to 0 and solve for x. Those are the x-intercepts. The polynomial can only intercept the x-axis at its intercepts. Therefore, given any interval between two x-intercepts, the points of the graph in that interval must be all positive or all negative (the graph cannot go from positive to negative without crossing the x-axis, since it is continuous). Test one point between each pair of intercepts to see if the graph is positive or negative. Then just fill in the graph. This will be a rough graph only, but, it is probably enough.

Here is an example. Graph $f(x) = x^3 - 4x$. The first step is to factor. The factored form of the function here is $f(x) = (x - 2)(x + 2)x$. The intercepts are at 2, -2, and 0, which are obtained by solving the factors for x, after setting them to 0. For example,

$x - 2 = 0$
$x = 2$

Now, plot (-2, 0), (0, 0), and (2, 0). The *y*-value is 0, because these are *x*-intercepts. Next, test points. Test anything less than -2. Let's pick -5. Putting -5 into the function, we get $(-5)^3 - 4(-5) = -105$. Since -105 is a negative number, the graph is negative everywhere $x < -2$. Now try a value between -2 and 0, say -1. Putting -1 into the equation, results in $(-1)^3 - 4(-1) = 3$, and so the graph is positive in that interval.

Next, try any number between 0 and 2, say 1. Putting 1 into the equation, we get $(1)^3 - 4(1) = -3$, which means the graph is negative in that interval. Finally, try any number greater than 2, say 10. $(10)^3 - 4(10) = 960$, meaning the graph is positive when *x* is greater than 2. Putting this altogether, the graph intercepts the *x*-axis at -2, 0, and 2. The graph is negative, positive, negative, and then positive (crossing at the intercepts). Thus, the graph looks just like Image 3.8.

Graphs of Rational Functions

Rational functions have breaks in their graphs, where the functions approach asymptotes. It is easy to recognize a rational function, as the algebraic formula for rational functions are fractions, with variables in the numerator and in the denominator. The first step in graphing them is to factor both the top and bottom and then cross out (cancel) any factors that are in both the top and bottom. Factors that remain in the bottom should each be set to 0 and solved. These solutions will be where asymptotes occur. Then it is just a matter of checking some points to see where the graph lies. Consider

$$f(x) = \frac{x^2 - 3x + 2}{x^3 - 2x^2 - 5x + 6}.$$

It is a rational function with variables (*x*'s) on top and bottom. The first step is to factor it. In factored form, the function is

$$f(x) = \frac{(x-1)(x-2)}{(x-1)(x-3)(x+2)}.$$

The common factor of factor $x - 1$ tells us that there is a hole in the graph at $x = 1$. To be fully correct, when graphing the function, put a hole at (1, 1/6). The 1/6 is found by plugging 1 into the function:

$$g(x) = \frac{(x-2)}{(x-3)(x+2)}.$$

Notice that *f*(*x*) and *g*(*x*) are almost identical functions. The function *g*(*x*) is *f*(*x*) with the common factor canceled. The graphs of *f*(*x*) and *g*(*x*) are almost the same. They only differ in a single point. In the graph of *f*(*x*), at (1, 1/6) there is a hole. In the graph of *g*(*x*), at (1, 1/6) there is not a hole.

Holes occur at *x*-values that cause the denominator to be 0 (it is an undefined mathematical operation to divide by 0; that is, don't do it!) and the division could be eliminated by canceling the denominator factor with a factor in the numerator. Asymptotes occur at *x*-values that cause the denominator to be 0 and the division cannot be eliminated by canceling the denominator factor with a factor in the numerator. The asymptotes of this function are at $x = 3$ and $x = -2$ (those are the spots at which x would cause division by 0 and those factors have not been canceled). See Image 3.9, where the dotted lines are asymptotes. The *x*-intercept is at $x = 2$ (find the *x*-intercepts by setting the numerator to 0).

Image 3.9: The dotted lines are the asymptotes.

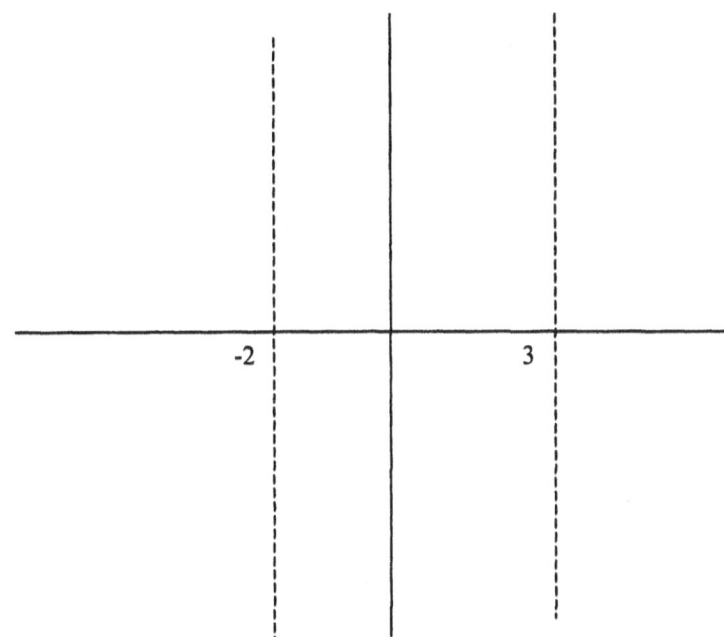

Now, we need to decide where the graph goes. For example, the graph might approach positive or negative infinity when it approaches the $x = -2$ asymptote. Thus, the graph looks like either Image 3.10 or Image 3.11. The graph is either below or above the *x*-axis (there are no *x*-intercepts there). Try any point, say -10. Putting -10 into the function, gives:
$$f(-10) = \frac{(-10-2)}{(-10-3)(-10+2)} = \frac{-12}{104},$$
which is a negative number. Thus, Image 3.11 is the correct one.

Chapter Three

Image 3.10: One possibility

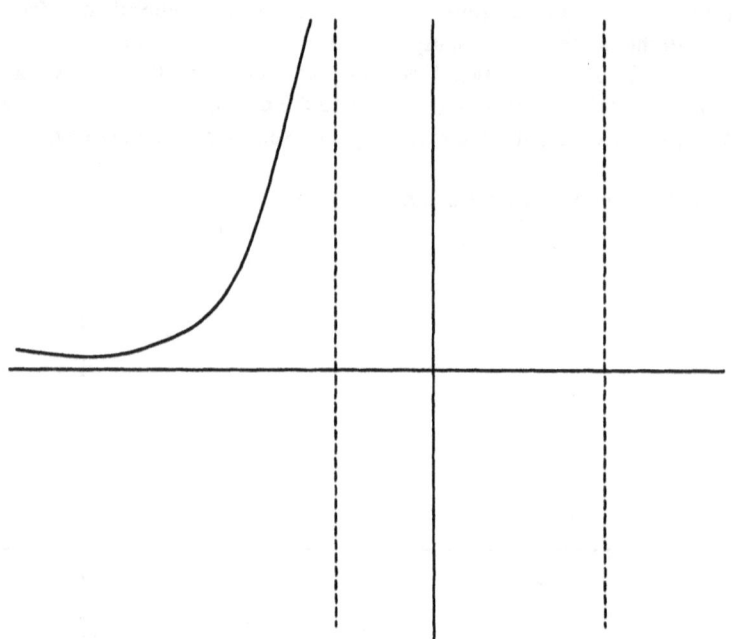

Image 3.11: Another possibility

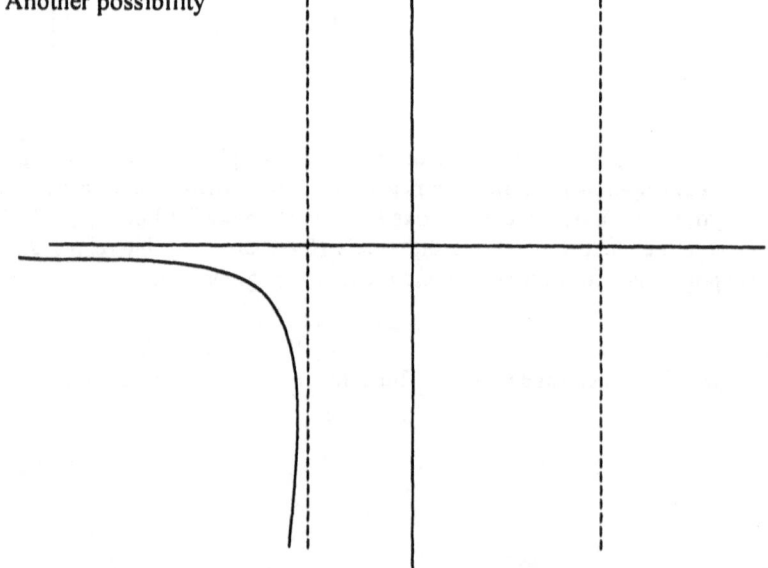

Graphing without a Graphing Calculator

Let's move on to the next part of the graph. Setting x to 0, there is a point on the graph at (0, 2/6). Also, recall that the x-intercept is at (2, 0). Therefore, the graph approaches the left asymptote and goes toward infinity. We do not know, however, if the graph crosses at (2, 0) and continues toward negative infinity (at the right asymptote) or turns at (2, 0) and continues toward positive infinity (at the right asymptote). But, this is easy to find out. Try any point between 2 and 3; say, 2.5. Plugging in 2.5 to the formula, results in:

$$f(2.5) = \frac{(2.5-2)}{(2.5-3)(2.5+2)} = -\frac{.5}{2.25}.$$

Since this is negative, the graph must cross at (2, 0) and continue toward negative infinity. Thus, Image 3.12 is what we have so far.

Image 3.12: Our answer so far

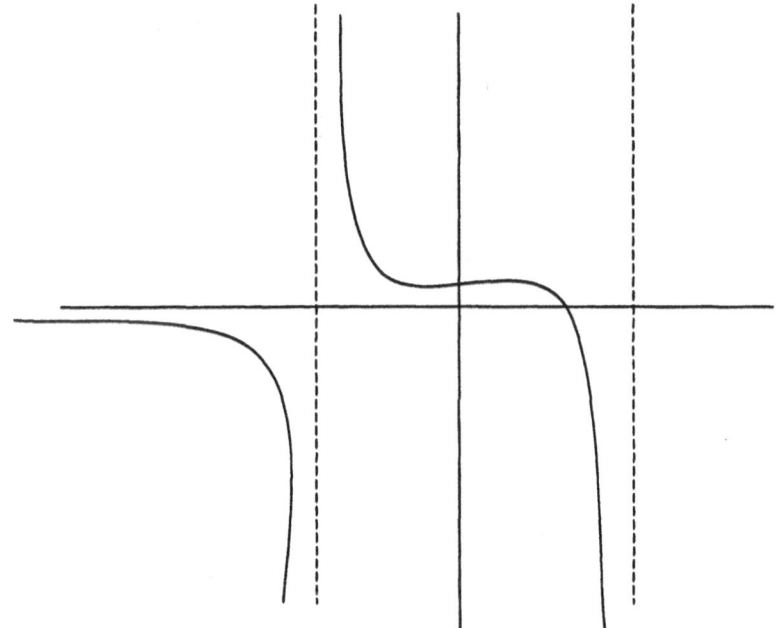

There is one more interval to figure out, where $x > 3$. There are no x-intercepts in this area. The graph must be completely negative or completely positive. Test a point to decide. Let's try $x = 5$. Plugging this into the equation, we get:

Chapter Three

$$f(5) = \frac{(5-2)}{(5-3)(5+2)} = \frac{3}{14}.$$

Since it is positive, our (almost) final graph looks like Image 3.13.

Image 3.13: Our (almost) final answer

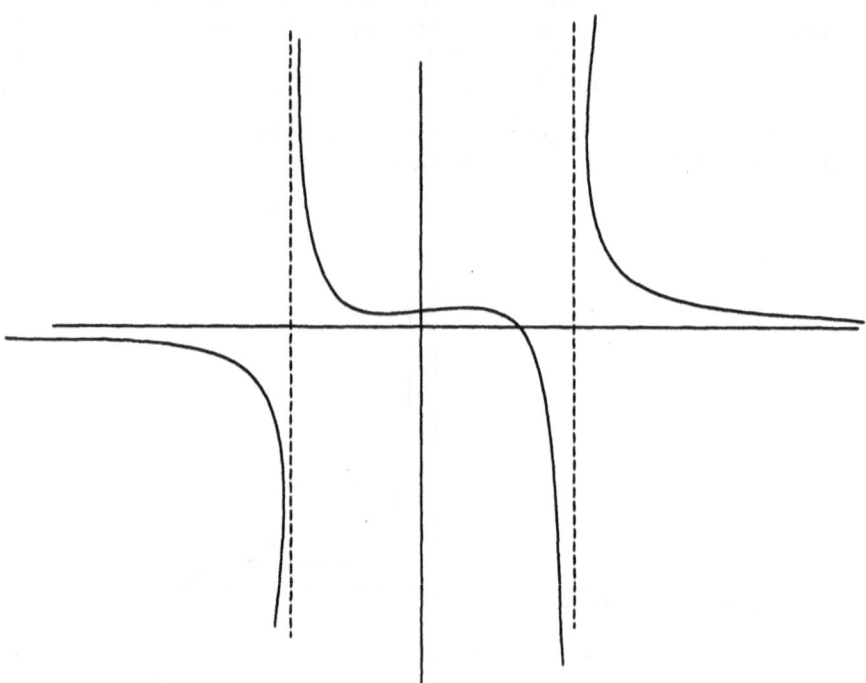

Image 3.13 is not the final graph because of one thing. Remember that hole at (1, 1/6)? That should appear in the graph. Draw a tiny open circle at that point. I have done so in Image 3.14. Now, the graph is perfect. Notice, though, that I haven't bothered to plot individual points. I haven't even plotted points we know, such as (5, 3/14). These points aren't really that important. We needed them to tell the general direction of the graph, and that is good enough for any purpose needed at the undergraduate level.

Image 3.14: Our final answer

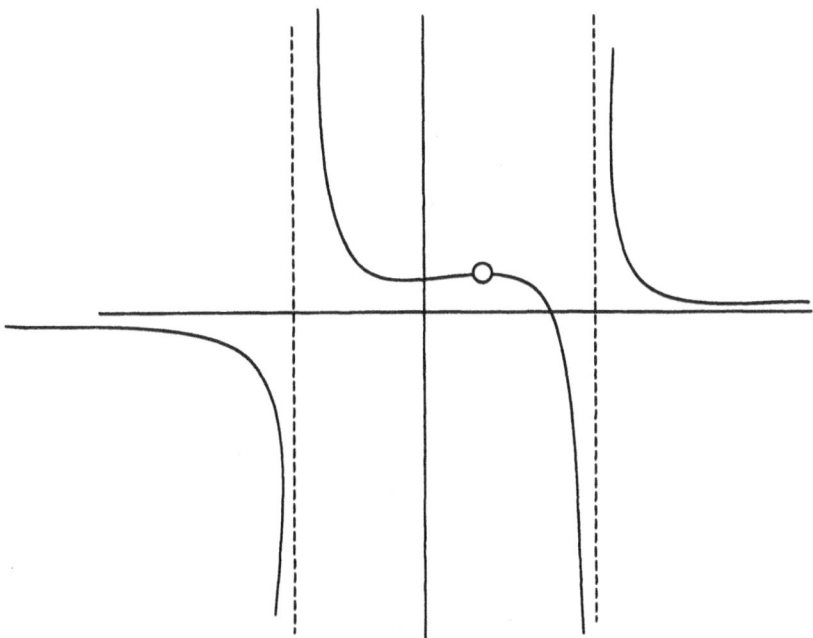

Graphs of Trigonometric Functions

Students should know that there are six trigonometric functions: sine (sin), cosine (cos), tangent (tan), cotangent (cot), secant (sec), and cosecant (csc). Images 3.15 and 3.16 show graphs of sine and cosine. Students should memorize these graphs.

Consider the functions $y = A\sin(Bx - C) + D$ and $y = A\cos(Bx - C) + D$, where $|A|$ is the amplitude. The period is determined by dividing 2π by $|B|$. Shifting left or right is called a phase shift. It is found by dividing C by $|B|$. The shift is left if C is negative and right if C is positive. D will shift the graph up (if D is positive) or down (if D is negative).

Let's say we need to graph $f(x) = 3\sin x$. The 3 is the amplitude. This is how high one hump of sine is. The distance from the largest to smallest value is twice the amplitude (or 6 in this case), but that uses two humps. See Image 3.17 and compare it to Image 3.15 (which is $\sin(x)$, not $3\sin(x)$).

Image 3.15: $f(x) = \sin(x)$

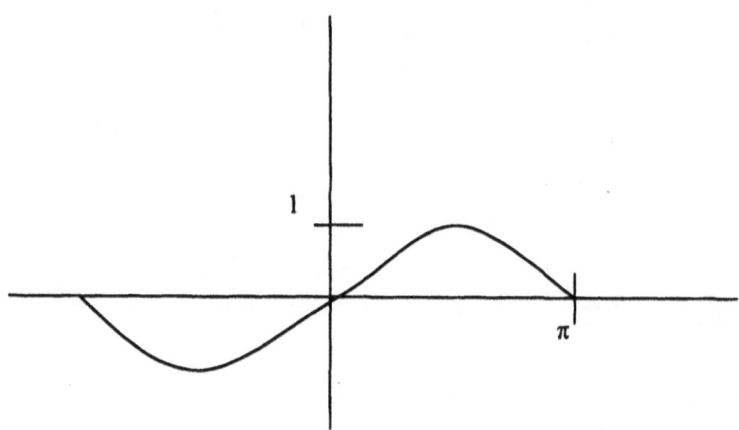

Image 3.16: $f(x) = \cos(x)$

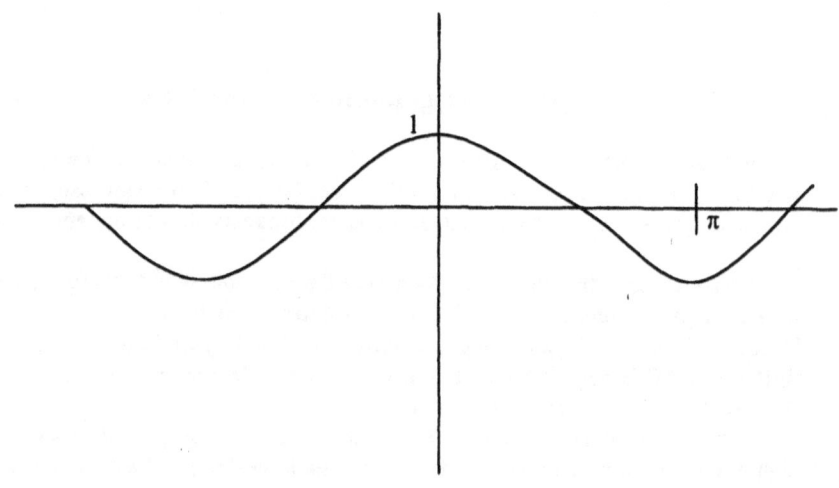

Image 3.17: $f(x) = 3\sin x$

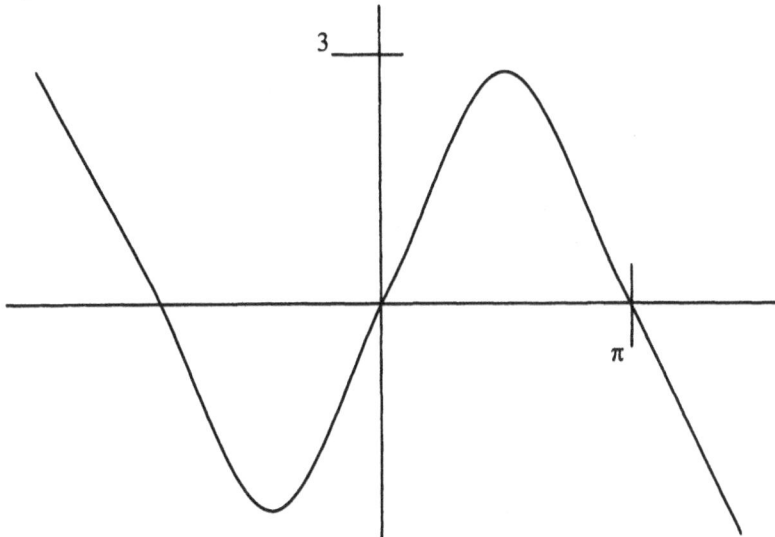

Now, let's say we need to graph $y = 3\sin(2x)$. The amplitude is 3, as before. But, the period has changed. The period is the width that it takes to get in one full cycle (or two humps, one up and one down). If the period is smaller than normal (remember that 2π is normal), then the graph will squish together. If the period is larger than normal, then the graph will spread wider apart. The period is found by dividing the regular period by B. In this case B = 2. The period is $.5\pi$. This will really squish up the function. See Image 3.18.

Compare this to Image 3.19, in which the function is spread out. The function, $f(x) = 3\sin(.5x)$, has a period of 2π divided by 1/2, or $2\pi(2) = 4\pi$.

Students can also use these facts to help in graphing:
   $\sin(-x) = -\sin(x)$
   $\cos(-x) = \cos(x)$
For example, when graphing $f(x) = -3\sin(-2x)$, it is easier to graph if one realizes that the negative with the -2 can pull out. Then, a negative times a negative is positive. The graph is equivalent to the graph of $f(x) = 3\sin(2x)$, which is easier to think about.

Image 3.18: $f(x) = 3\sin(2x)$

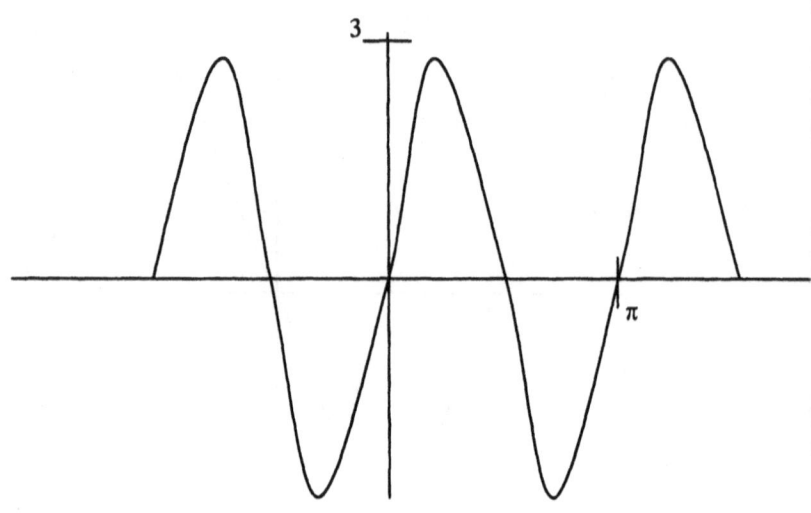

Image 3.19: $f(x) = 3\sin(\frac{1}{2}x)$

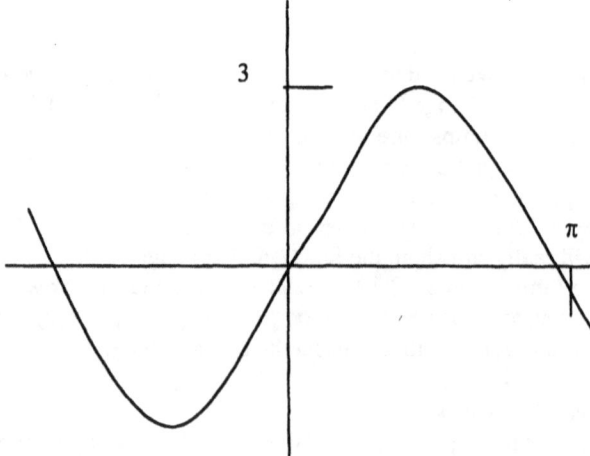

Chapter Four
A College Mathematics Placement Test

This chapter contains a full-length version of a typical college mathematics placement test. In fact, this is an actual placement test. Students should see a placement test for many reasons, not the least of which is to see their scores. It will also serve to help students know what skills they need to review and identify areas where assistance is needed. Having the ability to do each of the following skills is imperative for success in college mathematics. An incoming college student should be able to work problems involving:
- Order of operations
- Percent word problems
- Linear equations and inequalities
- Factoring polynomials
- Proportion word problems
- Multiplication of polynomials
- Simplifying radicals
- Evaluating functions (including composite functions)
- Quadratic equations
- Logarithmic and exponential equations
- Simplifying rational expressions
- Systems of equations
- Rational inequalities
- Geometric formulas, and
- Trigonometric concepts.

The placement test has items that test each of these skills. Students should use only one hour and no calculator when taking the placement test. Note that in this test all graphs are placed in x-y coordinate planes.

1. $-3 \cdot 4^2 - 16 - 4 + 3 - (1-2)^2 =$

      a. -162 b. -66
      c. -64 d. -42
      e. none of the above

2. $(9a^2 + 4a - 6)(4a) =$

 a. $36a^3 + 4a - 6$
 b. $9a^2 + 16a - 6$
 c. $9a^2 + 8a - 6$
 d. $36a^3 + 16a^2 - 24a$
 e. $9a^3 + 8a^2 - 6a$

3. Which of the following could be the graph of $10x - 2y = 20$?

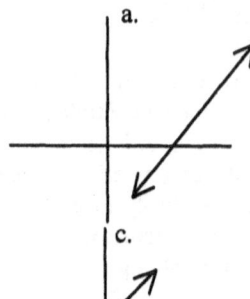
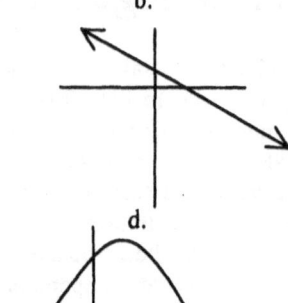
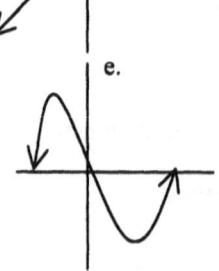
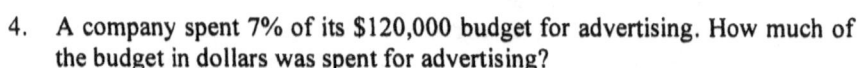

4. A company spent 7% of its $120,000 budget for advertising. How much of the budget in dollars was spent for advertising?

 a. $84
 b. $840
 c. $8,400
 d. $84,000
 e. $840,000

5. Which of the following is an equation that represents "four more than five times a number is six less than three times the number"?

 a. $4(5x) = 6(3x)$
 b. $5x - 6 = 3x + 4$
 c. $5x + 4 = 3x - 6$
 d. $4 + 5x - 6 = 3x$
 e. none of the above

A College Mathematics Placement Test

6. Solve for z. $5 - 2z = a$

   a. $\dfrac{5-a}{2}$ b. $\dfrac{5-a}{-2}$

   c. $\dfrac{a+5}{2}$ d. $-2(a+5)$

   e. $-2(a-5)$

7. Simplify $\sqrt{192}$ to lowest terms.

   a. $4\sqrt{12}$ b. $8\sqrt{3}$
   c. $12\sqrt{3}$ d. $16\sqrt{3}$
   e. $64\sqrt{3}$

8. Solve for z. $-3z \leq 5 + 2z$

   a. $z \leq -1$ b. $z \geq -1$
   c. $z \leq 1$ d. $z \geq 1$
   e. $z \leq 0$

9. Factor. $x^2 - 8x - 9$

   a. $(x-3)(x-3)$ b. $(x-3)(x+3)$
   c. $(x-9)(x+1)$ d. $(x+9)(x-1)$
   e. $(x-8)(x-9)$

10. Vonna scored 75 goals in her soccer practice. If her success-to-failure rate is 5:4, how many times did she attempt a goal?

    a. 60   b. 93
    c. 135   d. 169
    e. none of the above

11. If $f(x) = x^2 - 3x + 9$, then find $f(2)$.

    a. 3   b. 5
    c. 8   d. 10
    e. none of the above

12. Which of the following could be a graph of $y = x^2 - 3$?

a.
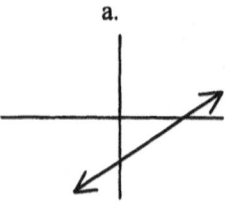

b.
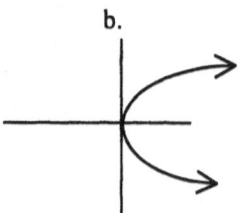

c.
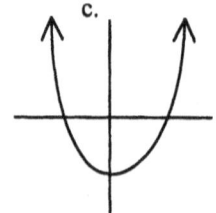

d.
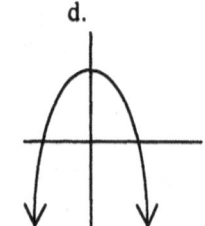

e.
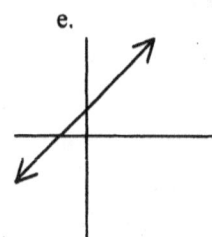

13. Given $7(x-2) = 4(x+1) - 21$, the solution for x lies between which of the following?

    a. –11 and -9   b. –2 and -.5
    c. -.5 and -.1   d. -.1 and .1
    e. none of the above

14. Given $x(4x - 11) = 3$, the sum of the solution set equals which of the following?

    a. $2\frac{3}{4}$ b. $5\frac{3}{4}$
    c. $-2\frac{3}{4}$ d. $3\frac{1}{4}$
    e. none of the above

A College Mathematics Placement Test

15. Solve for x. $2^x = 5$

    a. 2.5
    b. $\dfrac{\log(2)}{\log(5)}$
    c. $\log(\dfrac{5}{2})$
    d. $\dfrac{\log(5)}{\log(2)}$
    e. none of the above

16. Which of the following is an equivalent expression for $\dfrac{1-\dfrac{2}{x}}{1-\dfrac{3}{x}+\dfrac{2}{x^2}}$?

    a. $\dfrac{x^2 - 2x}{x^2 - 3x + 2}$
    b. $\dfrac{x-2}{x^2 - 3x + 2}$
    c. $\dfrac{x - 2x}{x - 3x + 2}$
    d. $\dfrac{x^2 - 3x + 2}{x^2 - 2x}$
    e. none of the above

17. $\log_2 8 =$

    a. 2
    b. 3
    c. 8
    d. 64
    e. 256

18. Given $\begin{cases} 2x + 5y = -2 \\ 3x - 4y = 20 \end{cases}$, then which of the following equals $x - y$?

    a. -8
    b. 4
    c. 2
    d. -2
    e. none of the above

19. Solve for x. $ax^2 = b$

    a. $\pm\sqrt{\dfrac{b}{a}}$
    b. $\pm\dfrac{\sqrt{b}}{a}$

c. $\pm\dfrac{b}{a}$

d. $\pm\left(\dfrac{b}{a}\right)^2$

e. none of the above

20. Which of the following is not the graph of a function?

a.

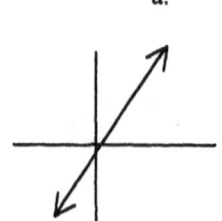

b.

c.

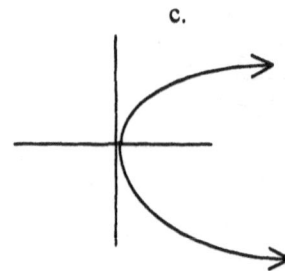

d.

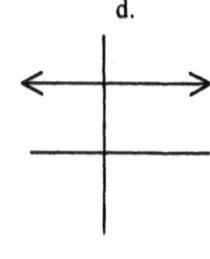

e. None of the above are graphs of functions.

21. Solve for x and answer with the sum of the solution set.

a. -9
b. 9
c. -3
d. 3
e. none of the above

22. Solve for x. $\log_{10}(3x^2) - \log_{10}(9x) = a$

a. 1
b. 10^a
c. 30^a
d. 3×10^a
e. $\log_{10}(10)$

A College Mathematics Placement Test

23. Solve for x. $e^{\ln x} = 2.5$

   a. 2.5
   b. ln(2.5)
   c. $e^{2.5x}$
   d. $e^{\ln(\frac{1}{25})}$
   e. none of the above

24. In which quadrant(s) are the values of $\sin\theta$ positive?

   a. I only
   b. I and IV
   c. I and II
   d. II and III
   e. II and IV

25. Give the center and radius of the circle whose equation is
   $(x - 2)^2 + (y - 4)^2 = 9$.

   a. center (-2, -4), radius 3
   b. center (2, 4), radius 3
   c. center (-2, -4), radius 9
   d. center (2, 4), radius 9
   e. none of the above

26. Solve for ß in the right triangle below.

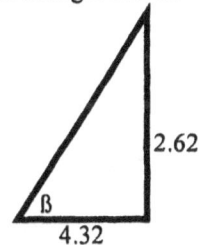

   a. $\tan\left(\dfrac{2.62}{4.32}\right)$
   b. $\tan^{-1}\left(\dfrac{4.32}{2.62}\right)$
   c. $\tan^{-1}\left(\dfrac{2.62}{4.32}\right)$
   d. $\sin\left(\dfrac{2.62}{4.32}\right)$
   e. $\sin^{-1}\left(\dfrac{2.62}{4.32}\right)$

27. What is the period of $y = A\cos(x - B\pi)$? (Here x is in radians.)

   a. $A\pi$
   b. $B\pi$
   c. 2π
   d. $2\pi / B$
   e. $A\pi / B$

28. Solve for x, in the right triangle below.

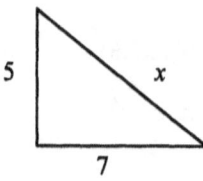

a. 12
c. 74
b. $\sqrt{74}$
d. 74^2
e. none of the above

29. What are the zeroes of the polynomial $x^3 - 2x^2 + x$?

    a. −1, 1
    c. −1, 0
    e. −1, 0, 1, 2
    b. 0, 1
    d. −1, 0, 1

30. Given $2x^2 + 7x = 4$, then find the sum of the solution set.

    a. -1
    c. $3\frac{1}{2}$
    e. none of the above
    b. −3
    d. $-3\frac{1}{2}$

Scoring

To score the test, first correct it. Here are the answers.

1. b
2. d
3. a
4. c
5. c
6. a
7. b
8. b
9. c
10. c
11. e
12. c
13. b
14. a

A College Mathematics Placement Test

15. d
16. a
17. b
18. e
19. a
20. c
21. c
22. d
23. a
24. c
25. b
26. c
27. c
28. b
29. b
30. d

Now, count the number correct on items 1 through 10. If the score is a 4 or less, then the student would place into remedial mathematics at the most basic level (a pre-algebra course). Such a course is often offered through a study skills department (not in the mathematics department). The course often counts for no university credit. Starting with such a course would place a student at least one and one-half years behind, even if that student only needs Calculus I (or an equivalent entry-level course). This low of a score would imply that the student is missing very elementary skills.

If the score is 5 or more on the first ten items, add up the points earned on items 11 through 20. A score of 4 or less on items 11 through 20 places the student into college algebra. This placement would put a student at least one year behind an entry-level course. College algebra might count as general credits, but usually does not count as mathematics credits.

If the score is 5 or more, calculate the score on the remaining ten items (items 21-30). A score of 4 or less results in placement into precalculus. This placement would place a student one-half of a year behind an entry-level course. A score of 5 or more means that the student would place into calculus (an entry-level course). This last placement is an excellent placement.

Chapter Five
Elaborate Solutions to the Test

The items are grouped by topic, and not only a solution to each item is given but further examples are offered. On a multiple-choice test, it is possible to get an item correct for reasons that do not really imply understanding the material. Of course, it is also possible to get an item wrong for a simple mistake. So, it is important to follow through the solutions on all items.

Item 1

This is the only item on order of operation. Whenever there is a string of operations (addition, subtraction, multiplication, and/or division), something must dictate what should be done first. Some people remember "Please Excuse My Dear Aunt Sally", because the correct order is Parentheses, Exponents, Multiplication, Division, Addition, and Subtraction. Actually, "Please Excuse My Dear Aunt Sally" is a tad misleading. Multiplication and division are equals, as are addition and subtraction. That is, multiplication is not necessarily computed before division. If there is both multiplication and division in a problem, then the one that occurs first, working left to right, takes precedence. Thus, multiplication and division are at the same level of precedence to each other and addition and subtraction are at the same level of precedence to each other.

Consider this sequence of operations: $3 - 4 \cdot 2$. The question is: Is the answer -2 (because $3 - 4 = -1$, and $-1 \cdot 2 = -2$) or -5 (since $4 \cdot 2 = 8$ and $3 - 8 = -5$). The order of operations says that the multiplication takes precedence over the subtraction. This makes -5 the correct answer and -2 wrong.

In Item 1, the parentheses are first according to the order of operations. Replace $-3 \cdot 4^2 - 16 - 4 + 3 - (1 - 2)^2$ with $-3 \cdot 4^2 - 16 - 4 + 3 - (-1)^2$, having taken care of the parentheses. Now, take care of the exponents, leaving $-3 \cdot 16 - 16 - 4 + 3 - 1$. Then, compute any multiplication and division, which results in $-48 - 16 - 4 + 3 - 1$. Finally, go through left to right doing the subtraction and/or addition. Leaving -66 as the answer.

Here are three more problems. $3 - 4 + 2 \cdot 5$. In this one, the $2 \cdot 5$ is done first (multiplication before addition or subtraction), which results in $3 - 4 + 10$.

Subtraction and addition rank equally, so work left to right and finish the problem. 3 - 4 = -1. And -1 + 10 = 9. The answer is 9.

6- 6 ÷ 2 Do the division first: 6 ÷ 2 = 3. Then 6 - 3 = 3.

3 • 7 - 9 ÷ 3 The multiplication and division are done before any subtracting, working left to right. Since 3 • 7 = 21 and 9 ÷ 3 = 3, the result is 21 – 3. Finally, 21 - 3 = 18, and 18 is the answer.

Item 2

This problem requires multiplying polynomials. Multiply the $4a$ by the $(9a^2 + 4a - 6)$, giving $36a^2 + 16a^2 - 24a$. The best thing to remember when trying to multiply polynomials is that everything in each term is multiplied by everything in each *other* term. $(a + b)(c + d) = ac + ad + bc + bd$, because a has to be multiplied by c and d. In addition, b has to be multiplied by c and d. Everything in the first term (a and b) has to be multiplied by everything in the second term (c and d).

Here is another situation: $(a + b)(c + d)(e + f)$. Take this two at a time: $(a + b)(ce + cf + de + df)$, having multiplied c by e and by f, and d by e and by f. Now, multiply a and b by everything in the remaining term, giving $ace + acf + ade + adf + bce + bcf + bde + bdf$.

Here is one using variables and numbers: $(7x + 5)(3x - 3)$. To solve this, multiply the $7x$ by $3x$ and by -3 and multiply the 5 by $3x$ and by -3. This results in $7x•3x - 7x•3 + 5•3x - 15$. Now, do the actual multiplication of the numbers and variables. x times x yields x^2. This results in $21x^2 - 21x + 15x - 15$, which equals $21x^2 - 6x - 15$, after combining the two "middle terms" (the two terms with a single x).

Here is one more: $(x + 3)(x - 2)(3x)$. Multiply the $(x - 2)$ by the $3x$ first. This results in $(3x^2 - 6x)$, since the x is multiplied by the $3x$ and the -2 is multiplied by the $3x$. So far we have this: $(x + 3)(x - 2)(3x) = (x + 3)(3x^2 - 6x)$. Now, multiply x (in the $x + 3$) by each of $3x^2$ and $6x$. Then multiply 3 by each of $3x^2$ and $6x$. This will leave: $3x^3 - 6x^2 + 9x^2 - 18x$. Finally, combine like terms. The answer is $3x^3 + 3x^2 - 18x$.

Item 3

The first step is to recognize the equation as that of a line. Further, answers d and e are not lines. In addition, the graph goes through the points (0, -10) and (2, 0). The only reasonable answer is a. Chapter Three gives further information about graphing.

Item 4

This is a percent word problem. To solve it, take .07 times 120,000. This will give answer c. Percent word problems will be easy to recognize, as the word percent or the percent symbol (%) will be used. Several forms of a percent word problems exist.

One form simply asks to find the percent of a number. For example, Marshall had approximately 85% of the items on a 250-item test correct. How many items did he have correct? We want to know what 85% of 250 is. To solve this, get rid of the percent sign by moving the decimal two spots to the left. Percent means per hundred. Moving the decimal one spot would be per ten. Moving the decimal two spots is per hundred. 85% = .85. Here are a few with answers given.

56% = .56
100% = 1.00 = 1
2% = .02
125% = 1.25
3% = .03

Once the decimal is moved, just multiply ("of" means multiply in mathematics). .85 • 250 = 212.50.

One more: What is 20% of 789? It is .20 • 789 = 157.8.

Let's consider a slight wrinkle in a percent word problem. An item costing $789 is on sale at 20% off. What is the sale price? We still need to know the 20%, but then subtract it from $789 to get the sale price. 20% of 789 is 157.80 (because .20 • 789 = 157.80). Then 789 - 157.80 = 631.20. Of course, we could have multiplied by .80 and then not subtracted. If it is 20% off, then we paid 80%. Let's do this: .80 • 789 = 631.20. Yes, we got the same answer.

Another form of a percent word problem gives us the result, and we are asked what the percent is. For example, let's say an item costing $789 is being sold for $750. We wonder what percent off that is. $789 - $750 = $39. What percent of $789 is $39? This time we don't know the percent. This is the opposite question of knowing the percent, so instead of multiplying we should divide. Changing "What percent of $789 is $39?" to math works this way: "What percent" is our unknown, say x. "Of" means multiply. So far, we have x • 789. "Is" means equals, now we have x • 789 = 39. To solve it, divide 39 by 789. The answer is approximately 5%.

Here is one more: Julius had 15 out of 35 questions correct on a test. What percent did he have correct? The equation is x • 35 = 15, as in what unknown (x) of (multiply) 35 is (=) 15. Dividing both sides by 35 gives us approximately .42857, so approximately 43%.

One last example: Carmen obtained a 2.5% raise and is now making $35,000 a year. What was her original salary? Many people work this incorrectly, because they take 2.5% of $35,000 and subtract it. This is wrong, because the 2.5% was of the original salary. Since the original salary was less than

$35,000, taking 2.5% of it would be a smaller amount than 2.5% of $35,000. Actually, we need to take 2.5% of an unknown number. We have this: original salary plus 2.5% of original salary is $35,000. Or $x + .025x = 35,000$. Or $1.025x = 35,000$. Then dividing 35,000 by 1.025, we will get the original salary, which is approximately $34,146.34.

Item 5

"Four more" means + 4, and "five times a number" is $5x$. So far, we have $5x + 4$. The word "is" means equals (=), now we have $5x + 4 =$. "Six less" means - 6, and "three times the number" is $3x$. At this point, we have $3x - 6$ on the other side of the equal sign. Together, this is $5x + 4 = 3x - 6$.

When converting English to mathematics, students should be aware of some key words that can be converted. For example, all of the following English words mean addition (+): sum, plus, more than, increased by, gain of, add, combined together, and total of. All of the following English words mean subtraction (-): minus, difference, less than, decreased by, less, subtract, and fewer than. All of the following English words mean multiplication (•): times, product, by, of, multiply, and factor of. All of the following English words mean division (÷): divide, quotient, separate equally, per, out of, ratio of, and into. The English word "double" means to multiply by 2 (•2) and the English words "half of" means to divide by 2 (÷2). All of the following English words can be replaced with the mathematical equal sign (=): is, are, was, were, will be, gives, yields, sold for, becomes, equals, amounts to, and is the same as.

Item 6 and Item 13

These two items involve solving linear equations. To solve the first one (Item 6), the a has to stay as an a. Since the problem requires solving for z, subtract the 5. This leaves $-2z = a - 5$. Now, divide by -2. The answer is $(a - 5)/-2$. Unfortunately, that is not a solution option. Multiplying both the top and bottom by -1 will give the answer $(-a + 5)/2$, which is still not an answer. However, we can switch the order of the 5 and the $-a$. This is answer a.

To solve the second one (Item 13), distribute the 7 on the left, and distribute the 4 on the right. This leaves $7x - 14 = 4x + 4 - 21$. Now, move the x's to the left, and the numbers to the right. Subtract the $4x$ and add on the 14. This results in $3x = 4 - 21 + 14$. Or $3x = -3$. Dividing by 3, we find out that $x = -1$. Thus, the solution lies between -2 and -.5, or answer b.

In general, to solve a linear equation move all terms with x's to the left, and move all numbers (without x's) to the right of the equal sign. One can move terms by adding, subtracting, multiplying, or dividing. Just be sure that what is done to one side of the equation is done to the other side of the equation. The goal is to have the equation solved for x, which means the equation looks like this: x = some number.

Here is another: $3x - 2 = 7x + 6$. Subtract $7x$ from each side. This leaves $-4x - 2 = 6$. Now add on 2 to each side, leaving $-4x = 8$. Finally, divide by -4. This means that $x = -2$. This is pretty much how solving linear equations work. Collect the x's and the numbers. To move either an x or a number, add or subtract it from both sides. If the original number (or x's) is added on (and you want to move it), then subtract it. If the original number (or x's) is subtracted (and you want to move it), then add it. This will always result in: $Ax = B$. Then, divide by A, and the answer is B divided by A.

In one situation this last step can be confusing, and that is when A is a fraction. For example, say we have: $(3/4)x = 8$. It easiest to multiply each side by $(4/3)$. Think of the last step as dividing by A, unless A is a fraction, and then multiply both sides by A's reciprocal (flip it over, the top becomes the bottom and the bottom becomes the top).

Item 7

To solve this problem, change 192 into a product, with one of the factors being a perfect square. For example, 192 equals 64 times 3. And the square root of 64 is 8. So,
$$\sqrt{192} = \sqrt{64 \cdot 3} = \sqrt{64} \cdot \sqrt{3} = 8\sqrt{3},$$
or answer b.

When asked to simplify a radical, the best rule to remember is:
$$\sqrt{ab} = \sqrt{a}\sqrt{b}.$$

Also memorize some of the square roots, and then those can be replaced with a number. The square roots that should be memorized include:

$$\sqrt{1} = 1 \quad \sqrt{4} = 2 \quad \sqrt{9} = 3 \quad \sqrt{16} = 4 \quad \sqrt{25} = 5$$
$$\sqrt{36} = 6 \quad \sqrt{49} = 7 \quad \sqrt{64} = 8 \quad \sqrt{81} = 9 \quad \sqrt{100} = 10.$$

These are called perfect squares. Perfect squares are numbers whose square roots are whole numbers. Of course, there are many more perfect squares than the ones I have listed (there are in fact an infinite number). The key when simplifying is to look for these square roots that result in whole numbers. For example,
$$\sqrt{200} = \sqrt{100 \cdot 2} = \sqrt{100} \cdot \sqrt{2} = 10\sqrt{2}.$$
It is fully simplified when there are no square roots that could be replaced by whole numbers. For example,
$$\sqrt{200} = \sqrt{4 \cdot 50} = \sqrt{4} \cdot \sqrt{50} = 2\sqrt{50}.$$
That is correct, but not fully simplified.
$$\sqrt{200} = 2\sqrt{50} = 2\sqrt{5 \cdot 10} = 2\sqrt{5}\sqrt{10}.$$
Again, correct, but not simplified.

$\sqrt{200} = 2\sqrt{50} = 2\sqrt{25 \cdot 2} = 2\sqrt{25}\sqrt{2} = 2 \cdot 5\sqrt{2} = 10\sqrt{2}$.

This time, it is both correct and simplified. To tell if a problem is simplified, you have to make sure that there are no more perfect squares under the radical.

Item 8

To solve this, move the $2z$ to the left, by subtracting it. This results in $-5z \leq 5$. Then, divide by -5, which flips the inequality sign. The answer is $z \geq -1$, or b.

Linear equations have an equal sign (=) and linear inequalities have a less than (<), less than or equal to ($\leq$), greater than (>), or greater than or equal to ($\geq$). To solve an inequality, treat it as if it were an equality. Solve just as if it had an equal sign, with one exception. When dividing or multiplying by a negative number, flip the inequality sign. If the inequality was <, flip it to >. If the inequality was $\leq$, flip it to $\geq$. If the inequality was >, flip it to <. If the inequality was $\geq$, flip it to $\leq$. That's it. Otherwise, pretend you have an equal sign and solve as you would for that.

Item 9

This problem involves factoring. That is, find a and b such that $x^2 - 8x - 9 = (x + a)(x + b)$. The product of a and b will be -9 and the sum of a and b will be -8. Three products are possible to give the -9: $1 \cdot -9$, $-1 \cdot 9$, or $3 \cdot -3$. Since the middle term is -8, use -9 and 1. Thus, the answer is c.

Factor $x^2 + 2x - 15$. Think about the products of -15. They are -1 and 15; 1 and -15; 3 and -5; and -3 and 5. Now think about which of those when added would give 2. Does $-1 + 15 = 2$? Of course not. Jumping ahead, $3 + -5 = -2$. Almost correct. $-3 + 5 = 2$. Thus $x^2 + 2x - 15 = (x - 3)(x + 5)$.

It is a little harder when the x^2 has a coefficient other than 1. Factor $6x^2 + 3x - 3$. We now have to worry about the 6, as well. But, the -3 at the end is easy because only -1 and 3, or 1 and -3 when multiplied give -3. For the six, 2 times 3 is 6, and 1 times 6 is 6, and those could be in either order. The best thing is to just try things. For example, $(6x - 1)(x + 3)$ yields the correct first and last term, but the middle term is $18x - x = 17x$. Switch the 6 and the 1. $(x - 1)(6x + 3)$ again yields the correct first and last term. The middle term is $-6x + 3x = -3x$. Almost correct. Try $(x + 1)(6x - 3)$, which again gives the correct first and last term. The middle term is $6x - 3x = 3x$. So, $(x + 1)(6x - 3)$ is correct.

It gets even a little harder when factoring a cubic. Factor out one term, by trying numbers and then reducing the cubic to a quadratic, and returning to previous methods for factoring that. To find the first root, try numbers. Numbers, when plugged in, will give 0 if they are roots. In addition, the roots have a special form. Given $ax^3 + bx^2 + cx + d$, the integer roots must be of the form $\pm(d/a)$. Try each of these (let each be x and plug in) to see if 0 results. Let's say the root is f. Then, the cubic equals $(x - f)$(a quadratic). To find the quadratic that is left divide the cubic by $x - f$, and then factor the quadratic.

For example, factor $x^3 - 3x^2 - 6x + 8$. The possible integer roots are -1, -2, -4, -8, 1, 2, 4, and 8. Since $a = 1$, the possible integer roots are just $\pm d$. Try some. $8^3 - 3(8)^2 - 6(8) + 8 \neq 0$, and thus 8 is not a root. $1^3 - 3(1)^2 - 6(1) + 8 = 0$, and thus 1 is a root. So, $(x - 1)$ is a term in the product, because 1 was a root. By the way if -1 was a root, then the factor would be $(x - -1)$ or $(x + 1)$. Now divide $x^3 - 3x^2 - 6x + 8$ by $x - 1$.

$$\begin{array}{r} x^2 - 2x - 8 \\ x-1 \overline{\smash{)}\, x^3 - 3x^2 - 6x + 8} \\ \underline{x^3 - x^2} \\ -2x^2 - 6x \\ \underline{-2x^2 + 2x} \\ -8x + 8 \\ \underline{-8x + 8} \\ 0 \end{array}$$

The thought process is basically this: How many times does x (in the $x - 1$) go into x^3? The answer is x^2. Then multiply x^2 by $(x - 1)$ to get $x^3 - x^2$. Subtract that from the polynomial. This leaves $-2x^2$. How many times does x (in the $x - 1$) go into $-2x^2$? The answer is $-2x$. Multiply $-2x$ by $x - 1$. This gives $-2x^2 + 2x$. Subtract that from the polynomial. Continue in this thought process. Polynomial division is just like long division. So, $x^3 - 3x^2 - 6x + 8 = (x - 1)(x^2 - 2x - 8)$. Now factor the remaining quadratic. Finally, $x^3 - 3x^2 - 6x + 8 = (x - 1)(x - 4)(x + 2)$.

Item 10

This is a proportion word problem. To solve it, solve this equation:

$$\frac{5}{4} = \frac{75}{x} = \frac{\text{success}}{\text{failure}}.$$

Thus $5x$ equals 75 times 4, or $5x = 300$, and $x = 60$. She had 60 failures. Then, 60 (the number of failures) plus 75 (the number of successes) is 135, which is the answer. This one is a little tough, because the problem asks for total times she attempted a goal (the sum of the successes and failures).

Proportion word problems are easily solved by using ratios set equal to each other and then cross multiplying. For example, say the ratio of girls to boys in a classroom is 4 to 5 and there are 20 boys in the classroom. Then how many girls are there? The set-up is this:

$$\frac{girls}{boys} = \frac{4}{5} = \frac{x}{20}$$

Using the words, girls and boys in this case, makes sure the number and variables are lined up correctly. We know there are 4 girls to 5 boys, so the 4 must be across from the girls and the 5 from the boys. There are 20 boys, so that needs to be on the boys' line.

To solve, cross multiply. In general that means:

$$\text{if } \frac{a}{b} = \frac{c}{d}, \text{ then } ad = bc.$$

I have multiplied the a by the d and the b by the c. I have cross multiplied.

In our case, 4 • 20 = 5 • x. Use an x for the unknown. Now solve it with 80 = 5x and 80/5 = 16. There are 16 girls in the classroom.

Item 11

This item is a function evaluation. Plug in 2 for x.

$$f(2) = 2^2 - 3 \cdot 2 + 9 = 4 - 6 + 9 = -2 + 9 = 7$$

The answer is e, none of the above. (It is likely that if "none of the above" is an option on a multiple-choice test that it will be used.) To evaluate a function is simply to plug in the given value for x and the result is y. For example, say $f(x) = 4x - 3$. Then, $f(7) = 4 \cdot 7 - 3 = 25$. We have evaluated the function at 7.

This item had nothing to do with composite functions, but composite functions are often on placement tests. Composite functions have the notation, $f \circ g(x)$, and mean $f(g(x))$. That is, the function $g(x)$ goes into the function $f(x)$. It is pronounced f of g of x. The key to a correct answer is to remember to work backwards from the equal sign. For example, if $f(x) = 2x$ and $g(x) = x + 3$, then $f \circ g(4) = f(g(4)) = f(7)$, since $g(4) = 4 + 3 = 7$, and then $f(7) = 2 \cdot 7 = 14$, and thus, $f \circ g(4) = 14$. Again, just work backwards from the equal sign. Using the same functions, $g \circ f(4) = g(8)$. This is true because $f(4) = 2 \cdot 4 = 8$, and then, $g(8) = 8 + 3 = 11$. So, $g \circ f(4) = 11$. Notice that function composition does not commute. We see an example of this with the fact that $f \circ g(4) \neq g \circ f(4)$.

Elaborate Solutions to the Test 53

This item also has nothing to do with function inverses, but that, too, is a topic that could appear on a placement test. To find the inverse of a function, exchange the y with the x and solve for y. Here is an example, find the inverse of $y = 3x + 4$. Exchange the y with the x to give $x = 3y + 4$. Now, solve for y. So, $x - 4 = 3y$, and then $y = (1/3)x - (4/3)$, which is the inverse of $y = 3x + 4$.

Finally, a word about the domain of functions. Another typical question is given a function, what is the domain. To answer, make the domain as large as possible, but exclude division by zero and negatives under a square root. So, the function $y = (x + 2)/(x - 3)$ has the domain of all real numbers except $x = 3$ (where there is division by 0). The domain of $y = \sqrt{x-3}$ is $x \geq 3$. In the last case, it is okay to have a 0 under a square root, but not a negative number.

Item 12

This item requires that the student recognize the quadratic function and what it looks like. Thus a and e are not reasonable options. It would also be helpful to recognize that this quadratic "opens up." The only reasonable answer is c. In addition, the parabola goes through (0, -3). Chapter Three gives further information about graphing.

Item 13

This item is explained with Item 6.

Item 14, Item 19, Item 21, and Item 30

All of these items involve solving quadratic equations. From the sheer quantity of items, students must realize that quadratic equations are really important. They will show up on placement tests, no doubt. Many methods for solving quadratic equations exist. The first, and perhaps easiest, is to factor the equation and set each factor equal to 0, and solve (just as if solving a linear equation). Of course, to factor you must have the equation in the form $ax^2 + bx + c = 0$. If the equation is not set to 0, then rearrange it until it is (that is, move everything to one side).

Another method is to use the quadratic formula. Whenever you can write a quadratic equation in this form $ax^2 + bx + c = 0$, you can use the quadratic formula. We will use it to solve Item 14. But, before we do, we must get the equation in Item 14 in the correct form. Distribute the x, which gives us $4x^2 - 11x = 3$. Then, subtract the 3, so that we have $4x^2 - 11x - 3 = 0$. Now we can use the

Chapter Five

quadratic formula. It is important that students memorize the quadratic formula. It is:

$$x = \frac{-b \pm \sqrt{b^2 - 4ac}}{2a},$$

where the a, b, and c are the coefficients as in $ax^2 + bx + c = 0$. If a student remembers the quadratic formula, then it is just a matter of plugging into the formula and one will have the solutions.

Doing this for Item 14, we have that $a = 4$, $b = -11$, and $c = -3$. Plugging in gives:

$$x = \frac{--11 \pm \sqrt{11^2 - 4 \cdot 4 \cdot -3}}{2 \cdot 4} = \frac{11 \pm \sqrt{121 + 48}}{8} = \frac{11 \pm \sqrt{169}}{8} = \frac{11 \pm 13}{8}.$$

And then separately,

$$x = \frac{11 + 13}{8} = \frac{24}{8} = 3, \text{ and } x = \frac{11 - 13}{8} = \frac{-2}{8} = \frac{-1}{4}.$$

The two solutions to this quadratic equation are 3 and -1/4. Adding these we get 2.75. (Normally, it is not necessary to add them. That is simply what the item wanted us to do.)

Item 30 will work in the same manner. First, move the 4 to the left, giving $2x^2 + 7x - 4 = 0$. Then, use the quadratic formula with $a = 2$, $b = 7$, and $c = -4$. Now plug into the formula.

$$x = \frac{-7 \pm \sqrt{7^2 - 4(2)(-4)}}{2(2)}$$

This results in roots of -4 and 1/2. Thus, the answer is d (once the roots are added together).

These two methods (factoring and quadratic formula) will be enough for most purposes. However, sometimes you have to do a few other things. To solve Item 19, simply divide by a, and then take the square root. This gives us answer a.

To solve Item 21, first square each side of the equation, in order to get rid of the square root. This results in $26 - 11x = (4 - x)^2$. Then, square out the right side, leaving $26 - 11x = 16 - 8x + x^2$. Move all variables and numbers to the same side of the equation. The result will be $0 = x^2 + 3x - 10$. Now, factor that, and solve for x. The factored form is $0 = (x + 5)(x - 2)$, and so the solutions are -5 and 2. Adding -5 and 2 gives -3. By the way, if the equation did not factor, then use the quadratic formula.

Elaborate Solutions to the Test 55

Item 15, Item 17, Item 22, and Item 23

These four items involve logs and/or exponential equations. A logarithm seeks an exponent. Chapter Two gives some basic rules on exponents. Consider $\log_3 9$. To evaluate that, think about what exponent will turn 3 into 9. From another view, $3^?=9$. The answer is 2, because 3 squared is 9. To solve Item 17, you need to understand that the logarithm means $2^x = 8$. Our job is to solve for x. This would be 3, since $2 \cdot 2 \cdot 2 = 8$.

The base (in our first example it is 3 and in Item 17, it is 2) can be any positive number, but is often 10 (in which case it is called the common log) or e (in which case it is called the natural log). The number e is precisely that, a number (similar to the number π). The number e is approximately 2.71828.

The answer to $\log_{10} 10$ has to be 1, because 10 to the power of 1 is 10. By the way, if you see a problem without a base (the little lower number is not given), assume the base is 10. Further, if the base is e, the log is written as ln (for natural log). That is, $\log_e x = \ln x$. Further, $\log_e e = 1$, because e to the 1 is e. And $\log_{10} 10^2 = 2$, because 10 squared gives 10 squared. Evaluate it to make this truth clear. Ten squared is 100, so $\log_{10} 100 = 2$. Further, consider $\log_{10} 10^x = x$. This one is a bit harder to think about, but remember that we need to know the exponent. $\log_{10} 10^x$ is asking $10^? = 10^x$. In Item 23, we use the fact that the exponential e and the natural log, $\ln x$, are inverses of each other. So, $e^{\ln x}$ is just x and then $x = 2.5$.

Students should memorize several rules about logs. They work with any base.

$$\log a + \log b = \log(ab)$$

$$\log a - \log b = \log(\frac{a}{b})$$

$$\log a^b = b \log a$$

These rules can be used to work with log equations. For example, to solve Item 15, remove the x from the exponent position. One method is to take the log of both sides, and then let the x fall down in front. This gives $x \log 2 = \log 5$. Then, divide by $\log 2$ to solve the equation for x. This gives the answer d.

To solve Item 22, use the fact that the subtraction can be turned into a division to give:

$$\log_{10}(\frac{3x^2}{9x}) = a.$$

And this is the same as:

$$\frac{3x^2}{9x} = 10^a.$$

We could try canceling to obtain:

$$\frac{x}{3} = 10^a.$$

Multiplying through by 3, we obtain answer d.

Item 16

Simplifying rational expressions is confusing to students because in expressions there is no equal sign. Therefore, one cannot multiply each term by the same number. For example: $3x + y$ is not the same rational expression as $6x + 2y$. Yet, if it was an equation such as $3x + y = 4$ then $6x + 2y = 8$ is an equivalent equation. The most common rule for equations (do the same thing to both sides of the equation) doesn't work in expressions because there is not an equal sign and thus not two sides to the equation.

The big concept to understand about rational expressions is to multiply by "fancy versions" of the number 1. For example, I might multiply by $3x/3x$. The purpose may be to obtain a common denominator. In Item 16, we need to multiply through by (x^2/x^2). In other words, multiply each term in the top and bottom by x^2. This gives us answer a, and we are already done.

Item 17

This item can be found with Item 15.

Item 18

This is a system of equations. We will cover two out of the several methods for solving systems of equations. Since either one will always work, two methods should be plenty. A system of equations is when there are more than one unknown and more than one equation.

Instead of starting with Item 18, we will start with an easier system. This one (as in Item 18) has two unknowns (x and y) and two equations.

$$\begin{cases} 2x + y = 3 \\ -x + y = 0 \end{cases}$$

One method is to add or subtract the two equations with the goal of eliminating (subtracting or adding out) one of the unknowns. Often directly adding or subtracting the two equations as they are given will not accomplish the elimination of one of the unknowns. In the case that it does not, first multiply one or both equations by some number and then add or subtract them. Also, I like to avoid subtracting (too confusing). In the example we have, I would rather add them. I will first multiply the second equation by -1. Remember that my goal is to eliminate one variable.

$$2x + y = 3$$
$$+\ x - y = 0$$
$$\overline{3x\ \ \ = 3}$$

Adding the equations (after multiplying the bottom one by -1) gives us $3x = 3$, because $y + -y = 0$. Since $3x = 3$, it must be that $x = 1$. Find y by plugging x into any given equation. $x - y = 0$, or $1 - y = 0$, so $y = 1$. We now have our solution $(1, 1)$.

Turning to Item 18, multiply the top equation by 3 and the bottom by -2. This results in:

$$\begin{cases} 6x + 15y = -6 \\ -6x + 8y = -40 \end{cases}.$$

Then, add the two equations together to get $23y = -46$. Solving this for y results in -2. Plugging -2 in for y in the top equation gives $6x - 30 = -6$. Solving this for x gives 4. Subtracting -2 from 4 gives 6. The answer is "none of the above," because the solution is $(4, 6)$.

Now, let's solve the first system by another method. The system was

$$\begin{cases} 2x + y = 3 \\ -x + y = 0 \end{cases}.$$

Eliminate one variable by taking either equation and solving it for a variable. For example, take $-x + y = 0$ and solve it for y. This gives $y = x$. Then substitute into the remaining equation. $2x + (x) = 3$, having substituted x for y in the top equation. Then $3x = 3$ and this is the same spot as in the above problem.

Let's also solve Item 18 through this previous method (it is called substitution). The system is:

$$\begin{cases} 2x + 5y = -2 \\ 3x - 4y = 20 \end{cases}.$$

Solve the top equation for x. This gives $2x = -2 - 5y$ and then $x = -1 - (5/2)y$. Now, substitute this new value of x into the bottom equation. This will give $3(-1 - (5/2)y) - 4y = 20$. Now solve this equation for y. $-3 - (15/2)y - 4y = 20$. And $-3 - (23/2)y = 20$. Then $-6 - 23y = 40$. Notice here that I multiplied everything by 2. I was tired of dealing with the fraction. Multiplying through by a number is fine with equations (not with expressions). Now, $-23y = 46$, and $y = -2$. Once we know y, we can substitute it into either equation to solve for x. This is as before.

Elimination (adding multiples of the two equations) or substitution (solving for one variable and substituting into the remaining equation) will always work with two equations and two unknowns. Usually, the solution is a single number answer. Sometimes, however, there is no solution. No solution occurs when the solution process leads to some false statement (such as $2 = 0$). The only other possibility is to get infinite solutions. Infinite solutions occur when the solution process leads to some true statement that is not dependent on x or y (such as $2 = 2$).

Item 19

This item is with Item 14.

Item 20

A function means that for any x, there is exactly one y answer. It is probably easiest to remember the vertical line test, which means, it is a function if you can drop a vertical line anywhere and only hit one point. Thus, they are all functions except for c.

Item 21

This item is with Item 14.

Item 22

This item is with Item 15.

Item 23

This item is with Item 15.

Item 24, Item 26, Item 27, and Item 28

These items are all in the category of trigonometric functions. To solve Item 27, one just needs to understand how to read the period from the standard equation. The standard equation is $y = A_1\cos(B_1 x - C_1\pi) + D_1$, where I have used subscripts so the reader isn't confused with the variables in the item. Further, the period is $(2\pi/B_1)$, so the answer is c, since B_1 is 1 in this item, thus, $(2\pi/B_1) = (2\pi/1) = 2\pi$.

Elaborate Solutions to the Test

For Item 24, one needs to know that sine is the *y*-value over the hypotenuse and we need to know in what quadrants *y*-values are positive. This is quadrants I and II, as the system is numbered this way:

```
   II  |  I
  -----+-----
   III | IV
```

In the first quadrant, both *x* and *y* are positive. In the second quadrant, *x* is negative and *y* is positive. In the third quadrant, both *x* and *y* are negative. And in the fourth quadrant, *x* is positive and *y* is negative.

Let's give all six trigonometric definitions, based on the right triangle below.

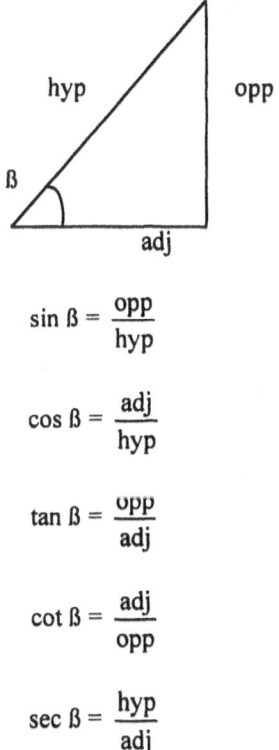

$$\sin \beta = \frac{\text{opp}}{\text{hyp}}$$

$$\cos \beta = \frac{\text{adj}}{\text{hyp}}$$

$$\tan \beta = \frac{\text{opp}}{\text{adj}}$$

$$\cot \beta = \frac{\text{adj}}{\text{opp}}$$

$$\sec \beta = \frac{\text{hyp}}{\text{adj}}$$

$$\csc \beta = \frac{\text{hyp}}{\text{opp}}$$

To figure out in which quadrants various trigonometric functions are positive, just view the triangle above as sitting in the x-y coordinate plane. View the opposite side as *y* and the adjacent side as *x*. Always think of the hypotenuse as being a positive value, and then let the *y* and *x* take on the sign of its quadrant. That is, *x* is negative in quadrants 2 and 3 and *y* is negative in quadrants 3 and 4. For example, say we want to know where cosine is positive. Cosine is adjacent over hypotenuse or *x* over hypotenuse. Since hypotenuse is positive, the question of interest is: where is *x* positive? In the first and fourth quadrant *x* is positive and, thus, that is where cosine is positive.

To solve Item 26, one has to understand some basic trigonometric identities given above. The easiest relationship to use in this item is the tangent one, and this would have 2.62 on top with 4.32 on the bottom. Since we are solving for the angle, it is the inverse function that we want. The answer is c. In other words, tan ß = (2.62/4.32), and then to solve for ß, take tangent inverse.

Item 28 uses Pythagorean Theorem: If *a* and *b* are the legs of a right triangle and *c* is the hypotenuse, then $c^2 = a^2 + b^2$.

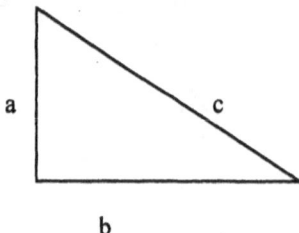

$5^2 + 7^2 = 74$, so the answer is $\sqrt{74}$. Pythagorean Theorem applies whenever one has a right triangle.

Although it might be good to have a complete course in trigonometry, I believe that a student can get away with knowing not very much! Students should know that cosine and sine are reciprocals, as are secant and cosecant, and cotangent and tangent. Students should know $\sin^2 x + \cos^2 x = 1$. Students should be able to convert from radians to degrees (multiply by 180° and divide by π) and from degrees to radians (divide by 180° and multiply by π). Students should probably memorize some basic trigonometric values, such as those that follow.

- $\sin(0) = 0$
- $\cos(0) = 1$
- $\tan(0) = 0$
- $\sin(30°) = 1/2$

- $\cos(30°) = \dfrac{\sqrt{3}}{2}$
- $\tan(30°) = \dfrac{1}{\sqrt{3}}$
- $\sin(45°) = \dfrac{1}{\sqrt{2}}$
- $\cos(45°) = \dfrac{1}{\sqrt{2}}$
- $\tan(45°) = 1$
- $\sin(60°) = \dfrac{\sqrt{3}}{2}$
- $\cos(60°) = \dfrac{1}{2}$
- $\tan(60°) = \sqrt{3}$
- $\sin(90°) = 1$
- $\cos(90°) = 0$
- $\tan(90°) =$ undefined
- $\sin(180°) = 0$
- $\cos(180°) = -1$
- $\tan(180°) = 0$

I don't expect my students to have memorized many identities for the trigonometric functions. Others may expect that the following identities are memorized:

$$\cos(x - y) = \cos x \cos y + \sin x \sin y$$

$$\cos(x + y) = \cos x \cos y - \sin x \sin y$$

$$\sin(x - y) = \sin x \cos y - \cos x \sin y$$

$$\sin(x + y) = \sin x \cos y + \cos x \sin y$$

$$\sin(2x) = 2\sin x \cos x$$

$$\cos(2x) = 2\cos^2 x - 1$$

$$\sin\dfrac{x}{2} = \pm\sqrt{\dfrac{1 - \cos x}{2}}$$

$$\cos\dfrac{x}{2} = \pm\sqrt{\dfrac{1 + \cos x}{2}}$$

$$\sin x \cos y = \frac{1}{2}[\sin(x+y) + \sin(x-y)]$$

$$\cos x \sin y = \frac{1}{2}[\sin(x+y) - \sin(x-y)]$$

$$\sin x \sin y = \frac{1}{2}[\cos(x-y) - \cos(x+y)]$$

$$\cos x \cos y = \frac{1}{2}[\cos(x+y) + \cos(x-y)]$$

Item 25

The general formula for a circle is $(x-a)^2 + (y-b)^2 = c^2$, where (a, b) is the center of the circle and c is the radius. So, the answer is b. The equation for a circle is one of many equations that students should probably memorize. I'll list the others here.

Rectangles and squares: The area of a square or rectangle is width times length: A = width•length, and the perimeter is the sum of the four sides. In a square, the width equals the length, but in a rectangle they may not be equal.

Parallelogram: The area of a parallelogram is base times height: A = base•height.

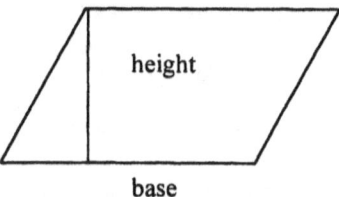

Triangle: The area of a triangle is one-half times base times height: A = (1/2)•base•height. To get the perimeter of a triangle, add the three sides.

Trapezoid: The area of a trapezoid is one-half times the sum of the two bases times the height: A = (1/2)•(top + bottom)•height.

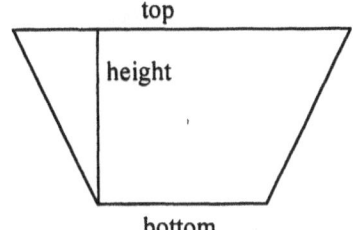

Circle: The area of a circle is π times the radius squared; that is, A = π(radius)². The circumference is 2 times π times the radius; that is, C = 2π(radius). An approximation for π is 3.14.

Box (cube): The volume of a box is the product of all three dimensions. The surface area can be found by summing the area of the sides (including the top and bottom).

Soup Can (a cylinder): The volume of a soup can is π times radius squared times height; that is, V = π(radius)²height. The surface area, not counting the top and bottom, is 2 times π times the radius times the height; that is, S = 2π(radius)(height). The total surface area, counting the top and bottom, is 2 times π times the radius times the sum of the radius and the height; that is, T = 2π(radius)(radius + height).

Ice-cream Cone: The volume of a cone is (1/3) times π times the radius squared times height; that is; V = (1/3)π(radius)²(height).

Ball (sphere): The volume of a ball is (4/3) times π times the radius cubed; that is, V = (4/3)π(radius)³.

Item 26

This item is with Item 24.

Item 27

This item is with Item 24.

Item 28

This item is with Item 24.

Item 29

To solve Item 29, you must understand that zeroes are the roots, or solutions. We could factor it, $x(x^2 - 2x + 1)$, and then $x(x - 1)(x - 1)$. And so the solutions (when each expression is set to 0) are 0, 1, and 1. This is answer b.

Item 30

This item is with Item 14.

An Item for Solving Rational Inequalities

There was not an item on the test that dealt with solving rational inequalities. I do not think the topic that important, but other mathematics professors do.

Here is an example: Solve for x. $\quad x^2 + x \leq 2$.
- a. $x \leq -2$
- b. $-2 \leq x$ or $x \geq 1$
- c. $-2 < x < 1$
- d. $-2 < x$ or $x > 1$
- e. $-2 \leq x \leq 1$

The answer is e. To solve it, first treat it like an equality and solve for x. So, subtract the 2, giving $x^2 + x - 2 \leq 0$, and then factor to give $(x - 1)(x + 2) \leq 0$. Solve each factor for x, which will give 1 and -2. At $x = 1$ and $x = -2$, the inequality $x^2 + x - 2 \leq 0$ is 0. The solution to this item is those x-values that cause the expression $x^2 + x - 2$ to be less than or equal to 0. The expression is 0 at 1 and -2, but what x-values cause the expression to be negative?

To find out, test each factor to see where it is positive and where it is negative. However, it is impossible to test every single point. It is also unnecessary to test every single point. Since the inequality is only 0 at $x = 1$ and $x = -2$, then x-values below -2 must all be positive or all negative (they cannot cross through 0 to change signs). In the same way, x-values between -2 and 1 must all be positive or all be negative. And x-values greater than 1 must all be positive or all be negative. It is only necessary to test one point in each of the intervals.

For x-values less than -2, try -3. $-3 - 1 = -4$, and so the factor $x - 1$ is negative for x-values less than -2. Putting -3 into $x + 2$, results in -1 and so the x-values are also negative for $x + 2$ for values less than -2. A similar process needs to be followed for x-values between -2 and 1, try 0. For $x - 1$, plugging in 0 gives -1 (again negative). For $x + 2$, $0 + 2$ gives 2, which is positive. Finally, try a value greater than 1, say 3. $3 - 1 = 2$, so the x-values are positive there. And $3 + 2 = 5$, so the x-values are positive there. The chart below contains a summary of the information.

Elaborate Solutions to the Test

	Less than −2	Between −2 and 1	Greater than 1
$x - 1$	− (negative)	−	+ (positive)
$x + 2$	−	+	+

Now, multiply out the negatives and the positives. For example, in the column below "less than -2", a negative times a negative is positive. In the next column, there is a negative times a positive, which is negative. Finally, in the third column, a positive times a positive is positive. See the chart below for this new information.

	Less than −2	Between −2 and 1	Greater than 1
$x - 1$	− (negative)	−	+ (positive)
$x + 2$	−	+	+
$(x - 1)(x + 2)$	+	−	+

Since the item requires x-values that are less than or equal to 0, the solutions are where the product is negative. So, the solution is all x-values between -2 and 1. Thus, the answer is either c or e. E is correct, because the solution must match the equal that was in the original problem.

Here is one more. Solve for x. $x^2 - x \geq 6$

First, move the 6 over to give $x^2 - x - 6 \geq 0$. Then factor, which gives $(x - 3)(x + 2) \geq 0$. Solving each factor, results in -2 and 3, which will create the intervals. In each of these intervals, test one point to see if it is positive or negative. The results are summarized below.

	Less than −2	Between −2 and 3	Greater than 3
$x - 3$	−	−	+
$x + 2$	−	+	+

Next, multiply out the positives and negatives. In the first column, two negatives multiplied give a positive. The next column is a negative times a positive, which is negative, and the final column is two positives which gives a positive. This information is summarized below.

	Less than −2	Between −2 and 3	Greater than 3
$x - 3$	−	−	+
$x + 2$	−	+	+
$(x - 3)(x + 2)$	+	−	+

Since the inequality was $\geq$, the solution is the positive areas. The solution is positive below -2 and greater than 3. So, $x \leq -2$ or $x \geq 3$.

Chapter Six
A Second College Placement Test with Solutions

This chapter provides another placement test. It is parallel to the one in Chapter Four. Parallel means that each item on one test is "identical" to an item on the other test (actually, it is the same numbered item). "Identical" means that the solution process is the same, however, the numbers and situations differ.

1. $2(4 + 1) - 2^3 + 6 \div 2 =$

        a. 3   b. 4
        c. 5   d. 7
        e. none of the above

2. $9x - 3[8 - 2(5 - 3x)] =$

    a. $3x - 34$ b. $3x + 34$
    c. $6 - 9x$ d. $9x + 6$
    e. $3x - 34$

3. Which of the following could be the graph of $y = x^2$?

a.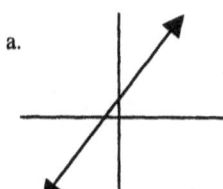
b.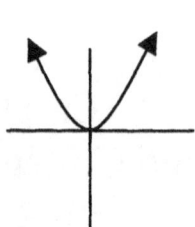

c.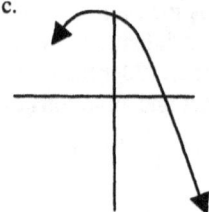
d.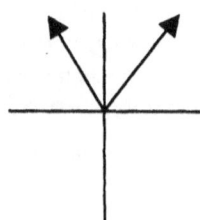

e.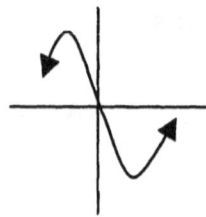

4. A department store has a blue blazer on sale for $114, which is 60% of the original price. What is the difference between the original price and the sale price?

    a. $45.60   b. $68.40
    c. $76   d. $114
    e. $190

5. Which of the following is an equation that represents "the quotient of r and the sum of r and four"?

a. $\dfrac{1}{r}+4$
b. $\dfrac{1}{r+4}$
c. $r(r+4)$
d. $\dfrac{r}{r+4}$
e. none of the above

6. Solve for z. $8z+6=a-2$

   a. $\dfrac{a-4}{8}$ b. $\dfrac{a-8}{8}$

   c. $a-1$ d. $\dfrac{a}{8}-8$

   e. $a(a+8)$

7. Simplify $\sqrt{200}$ to lowest terms.

   a. $10\sqrt{2}$ b. $100\sqrt{2}$

   c. $20\sqrt{10}$ d. $\sqrt{100}\sqrt{2}$

   e. $\sqrt{200}$

8. Solve for x. $3x < -2 - 2x$

   a. $x \le \dfrac{-2}{5}$ b. $x \ge \dfrac{-2}{5}$

   c. $x < \dfrac{-2}{5}$ d. $x \le -2$

   e. $x < -2$

9. Factor $x^2 - x - 20$

   a. $(x-4)(x-5)$ b. $(x-4)(x+5)$

   c. $(x-2)(x+10)$ d. $(x+4)(x-5)$

   e. $(x+4)(x+5)$

10. During a Friday, the ratio of stocks declining in price to those advancing was 5 to 3. If 450000 shares advanced, how many shares declined on that day?

    a. 270000   b. 300000

    c. 720000   d. 750000

    e. none of the above

11. If $f(x) = \dfrac{x^2 + 9}{3}$, then find $f(-3)$.

    a. 0   b. 3
    c. 6   d. 9
    e. none of the above

12. Which of the following could be a graph of $y = |\, x - 3 \,|$?

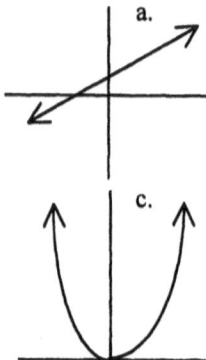
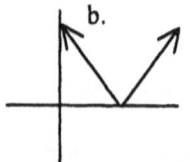
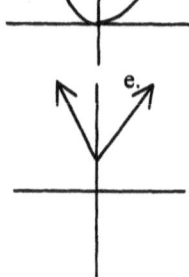
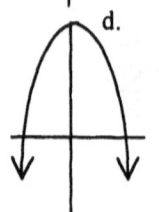

13. Solve for x. $(4x - 2)^2 = 8$.

    a. $\dfrac{1+\sqrt{2}}{2}, \dfrac{1-\sqrt{2}}{2}$ b. $\sqrt{8} + \dfrac{1}{2}$

    c. 18   d. $\sqrt{8} + \dfrac{1}{2}, \sqrt{8} - \dfrac{1}{2}$

    e. none of the above

A Second College Placement Test with Solutions   71

14. Given $4x^2 - 2x = 0$, then find the sum of the solution set.

    a. 0

    b. $\dfrac{1}{2}$

    c. 1

    d. $\dfrac{-1}{2}$

    e. none of the above

15. Solve for x. $2^x = 8^{x-4}$

    a. -3

    b. -1

    c. 2

    d. $\dfrac{\log(2)}{\log(8)}$

    e. none of the above.

16. Which of the following is an equivalent expression for
$$5x + \dfrac{4}{x} + \dfrac{5}{x^2} + 5x + 7x^2 + x?$$

    a. $11x + \dfrac{4x+5}{x+x^2} + 7x^2$

    b. $\dfrac{11x^3 + 4x + 5 + 7x^4}{x^2}$

    c. $11x^3 + 4x + 5 + 7x^4$

    d. $\dfrac{11x + 4x + 5x^2 + 7x^2}{x^2}$

    e. none of the above

17. $\log_2(\dfrac{1}{4}) =$

    a. –2

    b. 2

    c. 1/4

    d. -1/2

    e. 1/2

18. Given $\begin{cases} x+y=7 \\ 2x+3y=18 \end{cases}$, then which of the following equals $x - y$?

    a. 1

    b. 1

    c. 7

    d. 10

    e. none of the above

Chapter Six

19. Solve for x. $bx^2 + cx + a = 0$

   a. $\dfrac{-c \pm \sqrt{c^2 - 4ab}}{2b}$

   b. $\dfrac{-b \pm \sqrt{b^2 - 4ca}}{2a}$

   c. $\dfrac{c \pm \sqrt{a^2 + 4ab}}{2}$

   d. $\sqrt{\dfrac{c - 4ab}{2a}}$

   e. none of the above

20. Which of the following is not a function from the reals to the reals?

   a. $y = x^2$

   b. $y = 5x + 2$

   c. $y^2 = x$

   d. $5y = 4x$

   e. None of the above are functions.

21. Solve for x and answer with the sum of the solutions. $x + \sqrt{x - 4} = 4$

   a. 2

   b. 4

   c. 5

   d. 9

   e. none of the above

22. Solve for x. $\log_{10}(x + 3) + \log_{10}(x) = 1$

   a. –5, 2

   b. -5

   c. 1

   d. 2

   e. none of the above

23. Solve for x. $(\ln x)^2 = \ln x^2$

   a. $1, e^2$

   b. $\ln(x)$

   c. e^2

   d. $e^{\ln x}$

   e. none of the above

24. In which quadrant(s) are the values of tanß positive?

   a. I only

   b. I and IV

   c. I and III

   d. II and III

   e. II and IV

A Second College Placement Test with Solutions

25. Give the center and radius of the circle whose equation is
$(x + 5)^2 + (y - 2)^2 = 35$.

   a. center (-2, 5), radius 5
   b. center (5, -2), radius 35
   c. center (-5, 2), radius 7
   d. center (-5, 2), radius $\sqrt{35}$
   e. none of the above

26. Solve for c.

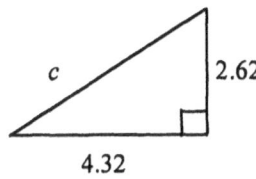

   a. $\tan(\frac{2.62}{4.32})$
   b. 5.05
   c. 6.94
   d. 10.41
   e. 25.53

27. Which of the following equations is a cosine function with an amplitude of 3 and a period of 4? (Here x is in radians.)

   a. $3\cos(\pi x/2)$
   b. $4\cos(3x)$
   c. $3\cos(4x)$
   d. $3\cos(\pi x/4)$
   e. none of the above

28. Solve for the angle β.

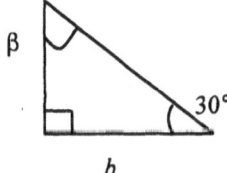

   a. $a + b$
   b. $\sqrt{a^2 + b^2}$
   c. 60°
   d. 240°
   e. none of the above

Chapter Six

29. Find all the vertical asymptotes of the graph of the rational function $f(x) = \dfrac{x+3}{x^2-9}$?

    a. $x = -3, x = 3$ b. $x = -3$
    c. $x = 3$ d. $y = -3$
    e. There are no vertical asymptotes.

30. Consider $25x^2 - 60x + 36 = 0$. Find the sum of the solutions.

    a. $\dfrac{6}{5}$ b. $\dfrac{5}{6}$
    c. $-\dfrac{6}{5}$ d. $\dfrac{6}{5}$ and $-\dfrac{6}{5}$
    e. none of the above

Scoring

To score this test, count the number correct in items 1 through 10 (including number 10). If there are 4 or less, the placement is in pre-algebra. If there are more than 4 correct, count the number correct in the items 11 through 20 (including numbers 11 and 20). If there are 4 or less, the placement is into college algebra. If there are more than 4 correct, count the number correct in the items 21 through 30 (including items 21 and 30). If there are 5 or less, the placement is into precalculus. If there are more than 5, the placement is into college calculus, or any other entry-level course.

1. c
2. c
3. b
4. c
5. d
6. b
7. a
8. c
9. d
10. d
11. c
12. b
13. a
14. e

15. b
16. a
17. a
18. a
19. c
20. b
21. b
22. d
23. a
24. c
25. d
26. b
27. a
28. c
29. c
30. e

Solutions are given for each item below. Remember that this test is identical to the previous test, so further examples of each item can be found in Chapter Five.

Item 1

This is an order of operations problem. Remember the "Please Excuse My Dear Aunt Sally" phrase. Take care of the parentheses first, which will leave $2 \cdot 5 - 2^3 + 6 \div 2$. Then, take care of the exponent, leaving $2 \cdot 5 - 8 + 6 \div 2$. Now, do the multiplying and dividing, from left to right, leaving $10 - 8 + 3$. Finally, do the subtracting and adding, from left to right. $2 + 3 = 5$. The answer is c.

Item 2

To solve this, distribute the -2 to leave $9x - 3[8 - 10 + 6x]$. Since it was a -2 that was distributed, the -2 times $-3x$ equals $+6x$. Then distribute the -3 to leave $9x - 24 + 30 - 18x$. Now, combine like terms to have $-9x + 6$. This is answer c (just placed in a different order).

Item 3

If one recognizes that the equation is a parabola, most of the choices are gone. Also, recognizing that the point (0, 0) is part of the curve allows one to choose the correct answer of b.

Item 4

This is a percent word problem. It is being sold for 60% of some unknown number. 60% of x is 114. Or, mathematically, $.60x = 114$. Dividing 114 by .60, the answer is 190. Then, 190 - 114 = 76, or c.

Item 5

Quotient is division, and sum is addition, so the answer is d. The quotient of r puts r on top, and on the bottom is the sum of r and 4.

Item 6

To solve this, move the 6 to the right by subtracting it. This gives $8z = a - 8$. Then, divide by 8 to give answer b. The a in this problem is just a dummy variable.

Item 7

To solve, write 200 as a product, using terms that might be perfect squares. For example, 2 times 100 is 200, and the square root of 100 is 10. Actually answer d is equal to the square root of 200, but it is not in lowest terms. Lowest terms replaces the square root of 100 with 10. Thus, answer a is correct.

Item 8

To solve this, just treat it like an equation, unless dividing or multiplying by a negative number. If that happens, then flip the inequality. First, add on $2x$ to each side to give $5x < -2$. Then, divide by 5. Since 5 is positive, there is no need to flip the inequality sign, and the answer is c.

Item 9

To factor, we need products of -20. These include 1 and -20; -1 and 20; -2 and 10; 2 and -10; -5 and 4; or 5 and -4. We also need the factors to differ by -1 (in order to get the -1 in front of the x term). The only choice then is -5 and 4 (since its product is -20 and their sum is -1). We want answer d.

Item 10

This is a proportion problem. The ratio of stocks declining to advancing is 5 to 3, and 450000 advanced. Mathematically, this means:

$$\frac{\text{declining}}{\text{advancing}} = \frac{5}{3} = \frac{x}{450000}.$$

To solve, cross multiply to get: $5 \cdot 450000 = 3x$. Or $2250000 = 3x$. Now, divide 2250000 by 3 to give answer d.

Item 11

Just plug in.

$$f(x) = \frac{(-3)^2 + 9}{3},$$

which is 18 on top and 3 on the bottom. Since 18 by 3 is 6, the answer is c.

Item 12

Absolute value functions are V's. Answer c and d are parabolas, so they are out. It is possible that answer a forms a V, and we just cannot see it. But, the essential point is that at $x = 3$, the y-value is 0. The answer a is out, as well as e. We are left with the correct answer of b.

Item 13

To solve this, square out the left side of the equation, leaving $16x^2 - 16x + 4 = 8$. Now, subtract the 8 to leave $16x^2 - 16x - 4 = 0$. And now, use the quadratic equation to solve for x. The quadratic formula is:

$$x = \frac{-b \pm \sqrt{b^2 - 4ac}}{2a},$$

with $a = 16$, $b = -16$ and $c = -4$. Plugging in,

$$x = \frac{-b \pm \sqrt{b^2 - 4ac}}{2a}$$

$$x = \frac{--16 \pm \sqrt{(-16)^2 - 4(16)(-4)}}{2(16)}$$

$$x = \frac{16 \pm \sqrt{256 + 256}}{32}$$

$$x = \frac{16 \pm \sqrt{512}}{32}$$

$$x = \frac{16 \pm \sqrt{2 \cdot 16 \cdot 16}}{32}$$

$$x = \frac{16 \pm 4 \cdot 4\sqrt{2}}{32}$$

$$x = \frac{16 \pm 16\sqrt{2}}{32},$$

which is answer "a" once it is simplified.

Item 14

To solve this, factor. This will give $2x(2x - 1) = 0$. Now, set each piece to 0. This results in $2x = 0$, giving $x = 0$. And $2x - 1 = 0$, giving $2x = 1$, and $x = 1/2$. Now, add the two together, $0 + (1/2) = 1/2$. The answer is b.

Item 15

This is a logarithm, but is easiest handled by switching the 8 to a 2. That is, $2^3 = 8$, so $2^x = (2^3)^{x-4}$. The 3 can be multiplied in, leaving $2^x = 2^{3x-12}$. Since the 2's are now equal (and not a 2 and a 8), they can be ignored. This leaves $x = 3x - 12$. Solve this for x, which results in $-2x = -12$, and $x = 6$, which of course is none of the above on the item.

Item 16

Since this is an expression (and not an equation), there isn't much to do, except combine like terms. Trying that gives:
$$11x + \frac{4}{x} + \frac{5}{x^2} + 7x^2.$$
The next thing is to examine the answers. It appears that a common denominator is sought in a few cases. The common denominator is x^2 (it is the highest power in any denominator and denominators can increase but not decrease). Each term must be multiplied by some form of the number 1 to change the denominator to x^2.

The first term gets multiplied by x^2/x^2, because its denominator was 1 and thus it needs the entire x^2 as a denominator. That term becomes $11x^3/x^2$. The $4/x$ gets multiplied by x/x, because it already had one x in its denominator. It becomes $4x/x^2$. The $5/x^2$ goes unchanged, because it already has the required denominator. Finally, the $7x^2$ gets multiplied by x^2/x^2 and becomes $7x^4/x^2$. Putting this altogether gives:
$$\frac{11x^3 + 4x + 5 + 7x^4}{x^2},$$
which is answer b.

Item 17

This is asking for x in this equation: $2^x = 1/4$. Since 4 can be written as 2^2, and since 4 is in the denominator, use a negative. That is: $1/4 = 2^{-2}$. Thus, $2^x = 2^{-2}$, which makes it obvious that $x = -2$.

Item 18

Although there are many ways to solve this, one method is to multiply the top equation by -2 and then add the two equations together. This results in:
$$\begin{cases} -2x - 2y = -14 \\ 2x + 3y = 18 \end{cases},$$

which results in $y = 4$, when the two equations are added together. If $y = 4$, then $x = 3$ (using $x + y = 7$, for example). And so, $3 - 4 = -1$, or a.

Item 19

This is a tricky problem in that it requires that one knows the quadratic formula, but then also rearranges it. The quadratic formula normally has the a, b, and c in different places. The regular quadratic formula is:

$$x = \frac{-b_1 \pm \sqrt{b_1^2 - 4a_1c_1}}{2a_1},$$

but that is when the quadratic is in this order, $y = a_1x^2 + b_1x + c_1$. I've used subscripts of the number 1 to tell the difference between the a, b, and c's used in the item's equation. In the quadratic formula, replace a_1 with b. So,

$$x = \frac{-b_1 \pm \sqrt{b_1^2 - 4bc_1}}{2b}.$$

Now, replace b_1 with c. So,

$$x = \frac{-c \pm \sqrt{c^2 - 4bc_1}}{2b}.$$

Finally, replace c_1 with a. This leaves

$$x = \frac{-c \pm \sqrt{c^2 - 4ba}}{2b},$$

which is answer a.

Item 20

Functions have to satisfy the idea that when one x goes in, one y comes out. In other words, there can only be one answer for each x that is plugged in. This is true for all of the equations except c. In c, consider, for example, putting a 4 in for x. Then, it is possible that $y = 2$, and it is possible that $y = -2$. That is two different answers. With functions, we always substitute in for x and get one answer for y. Based on that, c is not a function.

Item 21

To solve this, get rid of the square root. One idea is to move the x that is on the left to the right, and then square both sides. That is:

$$x + \sqrt{x-4} = 4$$
$$\sqrt{x-4} = 4 - x$$
$$x - 4 = (4-x)^2$$

Now square it out and gather like terms:
$$x - 4 = 16 - 8x + x^2$$
$$0 = 20 - 9x + x^2$$

Now we will factor:
$$(x-4)(x-5) = 0$$

And so, x is 4 and 5. However, when working quadratics in this manner (getting rid of the square root), sometimes roots that simply don't work are created. It is important to check those answers to see if they work. In this case, check the answers of 4 and 5. The 4 is fine, but the 5 doesn't work. See below.

$$4 + \sqrt{4-4} = 4$$
$$5 + \sqrt{5-4} \ne 4$$

So, only the 4 works, and the answer is b.

Item 22

To solve this, use the rules of logs. For example, the addition sign can be replaced by a single log, and then multiply. In other words $\log(x) + \log(y) = \log(xy)$. In this item, $\log_{10}((x+3)(x)) = 1$. Then multiply to get $\log_{10}(x^2 + 3x) = 1$. Since the answer is the exponent in logs, and an exponent of 1 does nothing, it is true that $10^1 = x^2 + 3x$, or $10 = x^2 + 3x$, or $x^2 + 3x - 10 = 0$. Now, factor, if possible. $(x+5)(x-2) = 0$. This means that $x = -5$ and $x = 2$. In logs, one should always check the answers and make sure they work. In fact, the $x = -5$ does not work, because it is not possible to put a negative inside a log, as in $\log(-5)$. The correct answer is 2, which is answer d.

Item 23

Knowing the log rules allows one to immediately drop the exponent down in front of the ln x^2 to give $2\ln x$. The other exponent means that there are two lnx's when multiplied by each other. So, we have $(\ln x)(\ln x) = 2\ln x$. Cancel one lnx from each side to give $\ln x = 2$. Then get rid of the natural log by using its inverse, e. So, $e^{\ln x} = e^2$ or $x = e^2$. A quick comment or two on inverses. Inverses are the opposite mathematical operation; they undo each other. For example, multiplication and division are inverses. Addition and subtraction are inverses. And lnx and e^x are inverses.

Now, it would seem that the answer here is c. But, it is not. When we divided by lnx, there is one case when one cannot divide by it, and that is when x

is 1, because $\ln(1) = 0$, and we cannot divide by 0. So, before answering c, see if 1 would also work. In fact, $(\ln 1)^2 = \ln 1^2$ does work, because $\ln(1) = 0$. So, it is just $(0)^2 = \ln(1)$, or $0 = 0$. So, the correct answer is a.

Item 24

Tangent is *y*-values over *x*-values. To be positive, either both *x* and *y* need to be positive or both need to be negative. That occurs in quadrants I and III.

Item 25

The equation for a circle is $(x - a)^2 + (y - b)^2 = r^2$, where (a, b) is the center of the circle and *r* is the radius. The center here is at (-5, 2) and the radius is the square root of 35.

Item 26

This is the Pythagorean Theorem. $4.32^2 + 2.62^2 \approx 25.5628$. The answer must be b, since we need to take the square root.

Item 27

Answers a, c, and d have amplitudes of 3. To have a period of 4, we need $(2\pi/B) = 4$, where B is the number that *x* gets multiplied by (the coefficient on *x*). This means that $B = (2\pi/4) = \pi/2$. The only one of a, c, and d (which all have the correct amplitude) with the correct B is a. Thus, the answer is a.

Item 28

The measures of the angles of a triangle adds to 180 degrees. This triangle has an angle with 30 degrees and an angle with 90 degrees (the right angle). Thus, subtract 30 and 90 from 180, so the answer is c.

Item 29

First, factor the bottom to see if there are any factors that the numerator and the denominator have in common.

$$f(x) = \frac{x+3}{x^2 - 9} = \frac{x+3}{(x-3)(x+3)}$$

Thus, there is a hole at -3, because $x + 3$ is a common factor. If one canceled the $x + 3$, it would leave $g(x) = 1/(x - 3)$. If one does cancel that factor out, it leaves a hole in that spot. The functions $f(x)$ and $g(x)$ are identical except at $x = -3$, where $f(x)$ has a hole and $g(x)$ does not. The factor that is not in both the numerator and denominator is where the asymptote is. So, there is an asymptote at $x = 3$ (from setting the factor to 0).

Item 30

It is probably easiest to use the quadratic formula to solve this. Recall that the quadratic formula is:

$$x = \frac{-b \pm \sqrt{b^2 - 4ac}}{2a},$$

and $a = 25$, $b = -60$ and $c = 36$.
Plugging in, gives

$$x = \frac{--60 \pm \sqrt{(-60)^2 - 4(25)36}}{2(25)},$$

which simplifies to

$$x = \frac{60 \pm \sqrt{3600 - 3600}}{50} = \frac{60}{50},$$

which appears to be answer a. However, actually answer a is incorrect. The reason is that 6/5 is a root twice. It is obtained when subtracting and when adding. Thus, the correct answer is (6/5) + (6/5), which is 12/5. The correct answer is e, none of the above.

Chapter Seven
Final Advice

Making the transition from high school mathematics to college mathematics has become a real bugaboo for students for many reasons. Even before secondary mathematics became reform, there was a transition issue. College is different from high school, no doubt. Many people view mathematics as a difficult subject. The combination of the transition issues with an already difficult subject (or at least a subject perceived to be difficult) has always made college mathematics a hurdle. However, when secondary mathematics reformed, this hurdle grew even larger. Secondary mathematics has changed in significant ways, and collegiate mathematics has not. Thus, the entering student has an even more difficult time, as now the average student has knowledge gaps.

To prepare for collegiate mathematics, it is important to do the following things. First, one needs to prepare mentally and emotionally. Accept the fact that college mathematics will be different from high school mathematics. Form a study plan ahead of time (see the appendix for information on how to form a study plan), and build up the emotional resolve it will take to stick to that study plan.

Second, once a student selects an undergraduate institution, check out the institution's web page, and search for information on placement testing. It is possible that the institution will provide a practice test for their placement test. Even if they do not, there should be information about the placement test. If there is not, students should contact the mathematics department or the college of science and engineering, and ask for information on the placement test. This is information that a university should give.

Third, a student should try to figure out what skills he or she is missing and find a method for learning the missing skills. This book, of course, is one such method.

Appendix One
Study Skills for College Mathematics

This book addresses the issue that students place incorrectly into remedial mathematics courses because they are not prepared for the placement test. But, even if a student makes it past the college mathematics placement test hurdle, it is still common to fail in mathematics. Students do not understand how to study in a college mathematics course. College mathematics courses differ from secondary mathematics courses in many manners. These include pedagogical differences (large class size, pace of the course, attitude of the professors) to content differences (an in-depth analysis approach with the procedures assumed already learned). This appendix will address some of the differences between high school and college and suggest what students might do to succeed.

Large Lectures

Often courses such as calculus are taught in large lecture with recitation format. Under this format, the professor will lecture to the entire class (the class could be as large as several hundred) on Monday, Wednesday, and Friday. On Tuesday and Thursday, smaller groups of perhaps twenty or twenty-five each meet with a graduate student to ask and receive answers for questions. Several features are problematic about this format.

The college professor can be unapproachable. College mathematics professors are not trained as teachers. They are trained as researchers. And, in fact, much of their job is doing research. Besides doing research, part of their job is called service, and it involves such things as committee work. Usually less than half their job is actually teaching. Combine this breakdown of their job with the fact that large lectures have hundreds of students, and one can begin to understand that college professors are often not aware of their students as individuals. It is unlikely that the professor will know students by name. The professor often is unable to tell whether the class is catching on to a topic and cannot field questions from the class during lecture.

In a large lecture format, students will have a graduate assistant, besides having a professor. Graduate assistants receive little or no training in how to be a teacher. They are likely to be top-notch students, but that does not always transfer to being good teachers.

Many graduate students in mathematics are international students. It is possible that the recitation they are leading is the first such experience they have had in United States. In fact, many arrive just days before their graduate program begins, and so their personal adjustments are enormous. They are trying to become acclimated to an entirely new culture (many mathematics graduate students are Chinese, and China is quite different from the United States). It is also possible that these international graduate students have some difficulty with speaking the English language, and their students will have to work extra hard just to understand them.

Pace of the Course

Even if students attend a small enough college that they are able to avoid the large lecture with recitation format, college mathematics classes move ahead at a very fast pace. Students do not work exercises during lecture time. The pace of collegiate courses is much quicker than high school, because a lot of material can be lectured on in fifty minutes. Once the professor lectures on a topic, he or she moves on to another topic. The student, in the meantime, must do the necessary things to truly learn the topic, as most students cannot learn a topic well enough for tests by simply attending lectures.

Independence of Students

The professor will expect students to be independent. Students need to accept that almost full responsibility lies with the student. The professor is responsible for presenting material. Many professors are not particularly good at presenting material, either. Professors do not feel responsible that students learn, and certainly do not feel that, if a student fails, it is the professor's fault. Professors do not even inform students what they need to do to succeed. Professors think that, too, is students' responsibility.

In college mathematics courses, then, to succeed, the student must understand and be able to work problems. The successful student does the things it takes to be able to do that: attends class, takes notes, reads the notes between lectures, does many problems, reads the textbook, and goes to office hours with specific questions. More details on these steps are given later in this appendix. The point here is that students must do these things without anyone reminding or suggesting that these things should be done.

In addition, the independence means that little to no time will be spent during the class time on group work projects. In NCTM-oriented curricula, much of the time is spent working in groups. It is expected in college mathematics that students work alone, both in class and out. Of course, students can form study groups on their own. But, in mathematics classes, think carefully about this. If a student is dependent on others to do the homework, tests are going to be even more of a challenge. In addition, if students express that they do the homework in groups to the professors, the professors will think of that as cheating.

Tests

Although it is very possible that a student has to work to earn money to pay the college and day-to-day bills, tests will fall when they fall. College professors do not adjust tests to fit students' schedules. Test dates might be given in class or announced in the syllabus.

Professors often do not review for tests. In a semester mathematics course, there may be one, two, or three tests and a final. This means that each of the tests covers an extensive amount of material. Material on tests might require the student to work problems that they have not seen in class, but should be able to do if they fully understand the material. It is not easy to receive an A letter grade. Points are not given for effort in college. Extra credit is seldom an option. In addition, students are seldom allowed to use a "cheat sheet," which is a common practice in high school. Students are not allowed to re-take a test in order to achieve a higher score, nor are there make-up exams given.

Attendance

Absence from class is treated very differently in high school than in college. In high school, students can expect to receive work ahead of time that they will miss and/or be allowed to make-up work when they return. Professors may or may not excuse absences, but even if they do, that has little meaning. Students in college are responsible for what they miss, and professors do not repeat or omit material for students who have been absent (even if the absence is for valid reasons). Professors often don't even tell the student what he or she missed. Students should be absent from college only when it is imperative, and then they must get notes from classmates. It is especially important not to miss a test, as a make-up test may not be given.

Office Hours

Office hours is a concept that does not occur in high school. Office hours are times that professors and graduate students promise they will be in their office. However professors do not just sit in their office hoping someone stops by. Professors are very busy with other obligations, and so students should make the best possible use of office hours.

When students are studying, they should keep a list of things that still are not understood. They should ask the recitation leader (usually a graduate student) to explain these things during the recitation time. If there is no recitation or recitation is not helpful, then the student can go to the recitation leader's office hours, or the professor's office hours. In courses with a recitation leader, students should go the professor's office hours only after having tried to get help from the recitation leader. If there is a tutoring center at the college, students can also go there.

It will upset a professor if a student simply shows up at office hours and says, "I don't understand anything." Even saying, "I don't understand Friday's lecture" is not a good thing to say. It is important to show evidence of having attempted to understand. One approach, for example, is to show the professor actual work on a certain problem, and then say something like, "I'm stuck right here. Could you give me a nudge forward?" The professor will see from this that the student is trying and should respond with help. Or, the student could show the professor the example that wasn't understood and say something very specific. "I see how you get from this step to this step, but I don't see how one would know to do that." General statements like, "I don't understand product mix word problems," will not be met with much help.

Study Plan

Faced with all these differences between high school and college, above all else, the successful student forms a study plan. In high school, it is possible that the student never had to study for a test. This is definitely not true in college. The study plan should include such things as always attending class and taking notes. Once the class is over and before the next class meets, the successful student reviews his or her notes and pays particular attention to things that he or she did not understand. The student should also read the textbook that covers the material on which the professor lectured.

When reading the textbook, the successful student has paper and pencil handy and will cover up the worked out part of the examples. The successful student then tries to work the examples without looking at the solution. Then the student will carefully compare his or her work to the solution. If correct, great. If not, the successful student will try to figure out why not. This process continues, as the student works through the reading in this manner, backing up and going back over examples that he or she got wrong, until convinced that he or she can work those problems correctly. If issues still occur, the student should make a list of things still not understood.

After reading lecture notes and reading the textbook, the successful student will attempt the odd-numbered exercises at the end of the section (this is done whether they are assigned or not). Usually the answers to odd-numbered problems are in the back of the book. It is also likely that the student can purchase a solution manual that actually shows all of the steps to the odd-numbered problems.

The successful student works these problems enough times that he or she not only can get the correct answer but can do so without outside aid (this includes other people, the textbook, or the calculator, if calculators are not allowed on tests). In addition, it is important to be able to work the problems rather quickly. After spending considerable time on the odd-numbered problems, the student should do any of the even-numbered problems that are actually assigned for homework.

Next, the successful student will actually memorize definitions, procedures, theorems, and other factual things that have been covered. Finally, he or she will briefly review all the previous sections that have not yet been tested.

All of these things should be done between each large lecture. It is imperative that the student does not fall behind. The students should plan to spend at least two to three hours on their mathematics class outside of class time, between each class session. In addition, about once a week, it is important to review the material that has already been tested, if there will be a cumulative final (which is likely).

The truth of the matter is that students do often fail at collegiate mathematics. But most of those students do not put the time and energy into studying as described above.

A Study

I did a study once in which I asked mathematics professors across my home state of Minnesota what the three top differences were between high school mathematics and college mathematics for students. I was amazed that all the professors gave the three same answers (of course, sometimes they used different words, and they weren't given in the same order). The three answers were:

Students must learn how to read a mathematics textbook.
Students in college are responsible for their own learning.
The pace is quicker and the depth of coverage deeper at college.

These three answers do not represent everything mentioned in this chapter, and, personally, I think issues such as class size prove to be a larger adjustment for students than the pace. Yet, the fact that all the professors contacted (one from each undergraduate mathematics department in Minnesota) mentioned these three concerns mean that they should be taken seriously.

Apparently, mathematics professors expect that students will read the textbook. In addition, the professors recognize that this is a skill that needs to be learned. Further, professors do not intend to cover everything that is needed and thus, students must read the textbook. This leads into the second issue, that students, not professors, are responsible for students' learning. Finally, the nature of having lectures (and not time in class to work on problems) versus how high school classes are structured means that material is covered quickly and deeply. Students are left to fill in the missing pieces.

Closing Thoughts

In sum, college and high school mathematics classes differ in many manners. These differences include large lectures in college and small classes in high school; faster pace in college; independent college students and more handholding in high school; stricter policies on tests (such as no rescheduling) in college; optional attendance in college; and the new beast that is office hours.

The issues discussed in this chapter will be present for students in NCTM-oriented curricula and in traditional curricula. Further, no method exists for preparing for these changes. It is not appropriate, for example, that high school teachers treat their students as if they were college students and expect the same level of maturity. However, freshmen should be aware of these differences and attempt to hit the ground running reacting to the changes. The quicker a freshman can develop a study plan the better off that freshman is.

Appendix Two
A Note to Teachers about State Standards

Secondary mathematics teachers (and administrators and curriculum specialists) are in a difficult position. They must prepare students for college. They also must meet state and federal standards. Unfortunately, the standards and college preparation are not as overlapping as they used to be. As a service to teachers, while writing this book, the author looked at the current standards of all fifty states. Any topic that was traditional in nature was included as part of this book, along with any topic that wasn't in the standards but is needed for college placement tests. Thus, this book can help teachers meet these standards.

Anything traditional is in this book, so a teacher could use the book as a supplement to a reform or NCTM-oriented curricula. Reform curricula emphasize processes over product. These processes include problem solving, communication, connections, representation, and technology. These processes are not included on placement tests. The only content that occurs in almost every state's standards that is not on college placement tests (and thus not reviewed in this book) is content related to data collection, probability, and statistics. Most college professors still desire to teach that material in college. However, it is (as mentioned) required by most state standards. Again, a teacher is safe because reform curricula contain content in this area. Thus, to meet state standards a teacher could adopt a reform curriculum and supplement it with this book.

Index

absolute value function, 16, 70n12, 77
amplitude, 31, 73n27, 82
approximation, 7
asymptote, 26-27, 74n29, 83
attendance, 89

ball, 63

circle, 41n25, 62, 63, 73n25, 82
college mathematics, 84
completing the square, 14-15
complex number, 8
composition, 52
cone, 63
cosine, graph of, 32
cross multiplying, 52
cube, 63
cubic function, 15
cylinder, 63

decimal, 6
degree, 60
denominator, 5
division by zero, 26
divisor, 4
domain, 53

elimination, 56-57
English words to math, 48
equation, 36n5, 48, 68n5, 76
exponent, 6, 35n1, 45
exponential decay, 18
exponential equation, 39n15, 39n19, 41n23, 55, 71n15, 81-82
exponential graph, 17
exponential growth, 17
expression, 39n16, 56, 71n16, 79

factoring, 37n9, 50-51, 69n9, 77
fraction, 5-6
function: definition of, 72n20, 80; evaluation of, 37n11, 52-53, 70n11, 77; graph of, 40n20, 58

graphing rules, 17-18

hole, 26, 83

ice-cream cone, 63
independence of students, 88
inequality, 37n8, 50, 64-65, 69n8, 76
inverse, 53

large lecture, 87
least common multiple, 4, 5
line: equation of, 37n6, 38n13, 46, 48-49, 69n6; graph of, 13, 36n3, 38n12, 46, 68n3, 70n12, 76
logarithmic function, 39n17, 40n22, 41n23, 55, 71n17, 72n22, 72n23, 78, 79, 81-82

matrix arithmetic, 10
matrix inverse, 11-12

National Council of Teachers of
    Mathematics (NCTM), 1
NCTM-oriented curricula, 1
NCTM standards, 1
negative number, 3
numerator, 5

office hours, 89
order of operation, 35n1, 45, 67n1,
    75

pace, 88
parabola: graph of, 14, 38n12, 53,
    70n12, 83; vertex of, 14, 77,
    78, 80. *See also* quadratic
    function
parallel lines, 14
parallelogram, 62
percent, 36n4, 47-48, 68n4, 76
period, 31, 33, 41n27, 58, 73n27,
    82
perpendicular lines, 14
please excuse my dear aunt sally,
    45, 75
polynomial: graph of, 24-26;
    multiplication of, 36n2, 46
polynomial long division, 51
prime number, 4
proportion, 27n10, 51-52, 69n10,
    77
Pythagorean theorem, 60, 82

quadrant, 59
quadratic formula, 53-54
quadratic function, 38n14, 42n30,
    53-54, 70n13, 71n14, 72n19,
    74n30, 77, 78, 80, 81. *See also*
    parabola

radian, 60
radical, 37n7, 40n21, 49, 69n7,
    72n21, 76, 80-81. *See also*
    square root function
ratio, 37n10, 51-52, 69n10, 77
rational function, 26
rational inequality, 64-65
rectangle, 62

sigma, 9-10
sine, graph of, 32
slope, 13
soup can, 63
sphere, 63
square, 62
square root, 8-9
square root function, 16. *See also*
    radical
state standards, 93
study plan, 90-91
substitution, 57
system of equations, solving of,
    39n18, 56-57, 71n18, 79-80

tests, 89
trapezoid, 62-63
triangle: area of, 62; perimeter of,
    62; solving for, 41n26, 42n28,
    73n26, 73n28, 82
trigonometric function, 31, 59-60,
    72n24, 73n27, 82; graphs of,
    31, 41n24
trigonometric identities, 33, 61-62

x-intercept, 27

zeroes, 42n29, 64, 67n2, 75

About the Author

CARMEN M. LATTERELL, Ph.D., is Associate Professor of Mathematics at the University of Minnesota Duluth. She has taught mathematics at all levels since 1988. She holds a bachelor's degree in mathematics from the College of St. Scholastica (Duluth, MN), a master's degree in mathematics from the University of Minnesota, and a Ph.D. degree in mathematics education from the University of Iowa. She is a frequent contributor to research journals on the subject of mathematics education, the author of *Math Wars: A Guide for Parents and Teachers* (Praeger, 2005), and the principal investigator on a large National Science Foundation grant to improve mathematics education.

www.ingramcontent.com/pod-product-compliance
Lightning Source LLC
Chambersburg PA
CBHW031714230426
43668CB00006B/206